T0328613

Developments in Environmental Modeling
Volume 31

Ecological Modeling: An Introduction to the Art and Science of Modeling Ecological Systems

Developments in Environmental Modeling

Developments in Environmental Modeling
Volume 31

Ecological Modeling: An Introduction to the Art and Science of Modeling Ecological Systems

Hsiao-Hsuan Wang
Ecological Systems Laboratory
Department of Wildlife and Fisheries Sciences
Texas A&M University, Texas, United States

William E. Grant
Ecological Systems Laboratory
Department of Wildlife and Fisheries Sciences
Texas A&M University, Texas, United States

ELSEVIER

Elsevier
Radarweg 29, PO Box 211, 1000 AE Amsterdam, Netherlands
The Boulevard, Langford Lane, Kidlington, Oxford OX5 1GB, United Kingdom
50 Hampshire Street, 5th Floor, Cambridge, MA 02139, United States

First edition 2019

ISBN: 978-0-444-64163-2
ISSN: 0167-8892

For information on all Elsevier publications
visit our web site at https://www.elsevier.com/books-and-journals

Publisher: Candice Janco
Acquisition Editor: Candice Janco
Editorial Project Manager: Laura Okidi
Production Project Manager: Omer Mukthar
Cover Designer: Christian Bilbow

Typeset by SPi Global, India

Working together
to grow libraries in
developing countries

www.elsevier.com • www.bookaid.org

Cover design concepts

"When I tried to understand the focus of this book, even though it demonstrated the advancements of technology, I could only think of flora, fauna, and alchemists in old antique handwritten books. Thus, I decided to have hand-drawn illustrations for the cover of this book and adapted the concept of "Theosophische Darstellung zur Alchemie, 1678." I kept the shapes of round, triangle, and heptagram, but replaced the alchemical emblems to scientific observations, experiments, records, results, and, of course, ideas!

Comparing to science with clear results, art could be interpreted through personal experiences and feelings. Therefore, you might not have the same feelings as mine when you see the cover and read the book. However, I do hope to deliver the beauty of art with science to you.

Finally, I would like to dedicate the cover to all scientific spirits and my magician sister."

Cover Designer
Rui-Pu Wang
Taipei, Taiwan

—

「在瞭解這本書的中心主旨時，雖然是日新月異的科技，腦中浮現的印象卻是泛黃的手抄古籍，裡頭畫有各種動植物、以及做著各種實驗的煉金術師。

於是我決定以手繪方式呈現，參考了煉金術的原理符號圖，保留了三角型、七芒星、圓，而現代科學的觀察、實驗、紀錄、成果、還有無論科學和文學都最需要的「靈光一閃」……等等，則一一取代原本的元素融入其中。

相對於有答案的科學，藝術通常是各自表述，各位在圖中讀到與我初衷有異的想法也無妨，只希望帶給各位的都是美好。

這幅畫獻給科學精神、和我的魔法師姊姊。」

封面設計
王瑞璞
台北、台灣

Contents

Preface

We have written this book with the intention of providing an introduction to ecological modeling that is both conceptually rigorous and easy-to-read. We envision its use as a textbook for graduate-level and upper-division undergraduate-level courses covering topics related to the use of systems analysis and simulation in ecology, natural resource management, and the environment. It also should serve as a primer and reference book for natural resource and environmental professionals at both managerial and policy levels whose jobs involve the development, evaluation, and/or use of ecological models.

As the title suggests, we view ecological modeling as both an art and a science. The science has its foundations in systems theory. The art has its foundations in our innate ability for abstract representation of the world around us. Thus by "art" we simply wish to convey the idea that there is more to the modeling process than can be explained based on scientific principles alone. We have tried to achieve an appropriate balance between the artistic and scientific aspects of the modeling process. There is an understandable tendency to become preoccupied, even in introductory courses, with the details of quantitative modeling techniques. However, in our experience, the most demanding intellectual activity associated with the modeling process is the development of the conceptual model. This is an art which must be learned through practice, there is no precise recipe to follow. But there are general scientific guidelines grounded in general systems theory. Not unlike the composition of music, which has no axiomatic basis in mathematics yet draws heavily on the mathematical principles of sound, development of conceptual models of ecological systems has no axiomatic basis in general systems theory yet draws heavily on systems principles. First and foremost, is the principle that system structure is the source of system behavior and system behavior, in turn, is the source system structure. Stated more simply, how a system changes from today to tomorrow depends on what it is like today and, depending on what happens today, the system is likely to be different tomorrow with regard to something that is important to us. The artistic aspects of representing (modeling) a series of such changes in a useful way are aided greatly, but not negated, by keeping this basic systems principle in mind.

Thus, a primary goal of this text is to provide insight into the "artistic" side of ecological modeling. Our belief is that one must master a relatively small set of quantitative techniques to begin to gain the working knowledge of modeling that comes only from first-hand experience. Use of sophisticated quantitative techniques should be postponed until the modeling process itself is well understood. But we hasten to add that gaining useful experience in the development of conceptual models often is facilitated by the feedback that can be provided by their quantification and execution. The unambiguous nature of numerical calculations most often identifies conceptual flaws that are undetectable based on diagrams and words alone. We have limited the math used in the body of the text to the simplest arithmetic capable of providing the feedback necessary to evaluate the conceptual models under consideration.

We have divided the text into four parts. In Part 1 (Chapters 1 and 2) we present the systems perspective. In Chapter 1, we consider why we should model ecological systems, introduce some basic terminology, compare simulation modeling to laboratory experiments and field observations, and present some basic characteristics of systems as well as some sources of surprising systems behavior. In Chapter 2, we describe basic building blocks of system structure which give rise to characteristic patterns of system behavior. In Part 2 (Chapters 3–7) we present the modeling process. In Chapter 3, we provide an overview of the four phases into which we have divided the modeling process, and present a few simple numerical examples. In Chapters 4–7, we describe conceptual model formulation (Chapter 4), quantitative model specification (Chapter 5), model evaluation (Chapter 6), and model application (Chapter 7) in greater detail. In Part 3 (Chapters 8 and 9), we present some practical considerations for modelers, including common pitfalls into which modelers often fall (Chapter 8), and a candid ("behind closed doors") view of the modeling process with some "guided" modeling exercises (Chapter 9). In Part 4, we reflect upon selected topics related to modeling ecological systems. In Chapter 10 we focus on the semantics, philosophical viewpoints, methodological approaches, and criteria associated with model validation (we prefer use of "evaluation" over "validation"). In Chapter 11, we compare systems modeling with other methods of problem-solving and consider the expectations we should, and should not, have of ecological models, both as modelers and as environmental decision makers. In Chapter 12, we consider the concept of "embracing" the uncertainty associated with ecological modeling, and offer brief summaries of uncertainty typologies, concepts, levels, and types. In Chapter 13, we comment upon the potential benefits and dangers of integrating large existing models, as opposed to developing models from "scratch." In Chapter 14, we reflect upon the commonalities among several alternative quantitative modeling frameworks from the systems perspective. Finally, in Chapter 15, we ponder the artistic aspects of modeling, particularly within the context of the science of ecology, that is, we offer some thoughts on the "art" in the art and science of ecological modeling.

The content of this book has evolved over the course of the four decades during which the elder of us has been teaching ecological modeling. A first text in the 1980s was motivated by the fact that most modeling texts available at the time were written almost exclusively for engineers. Ecological modeling books, including some landmark works (e.g., the four-volume series *Systems Analysis and Simulation in Ecology* by Bernie Patten, published in the 1970s), were primarily reference volumes. A second text in the 1990s was prompted by a reconsideration of some concepts involving model validation/evaluation and by the advent of the possibility to provide readers direct access to computerized versions of the models used the text. A third text in the 2000s resulted from the realization that presentation of nonessential quantitative details and computer code tended to divert attention from the conceptual foundations upon which the modeling process is based.

The present text was motivated by the concern that rapidly expanding access to user-friendly software supporting quantitative aspects of ecological modeling is being accompanied by increased misuse of these tools. In recent years, we have reviewed a growing number of theses and dissertations, government reports, and even manuscripts submitted to scientific journals such as *Ecological Modeling*, describing (unintentional) misuse of models. Just as rapid technological advances in cell phones arguably have not been accompanied by commensurate development of cell phone etiquette, rapid technological advances in systems modeling software arguably have not been accompanied by commensurate understanding among users of the systems perspective. Our hope is that the present text will provide readers with (1) a commonsense understanding of the systems perspective and its foundations in general systems theory, (2) an appreciation for the importance of qualitative aspects of the modeling process, (3) the ability to evaluate and use ecological models appropriately, and (4) the desire and confidence to embark upon the modeling experience. Above all, if the present text motivates readers to learn more about the art and science of modeling ecological systems, we will have accomplished our goal.

We would like to thank Brian Fath for his invitation to publish this book as part of the series titled *Developments in Environmental Modeling*, for which he is Editor-in-Chief. We also would like to thank Brian, as well as Volker Grimm, Takuya Iwanaga, and Alexey Voinov for reviewing parts of the book and for providing most helpful suggestions regarding its content. Finally, we would like to thank Sandra Marin, Ellen Pedersen, and Todd Swannack, all of whom have coauthored an ecological modeling textbook with the elder of us, for their ideas and insights into needs of beginning modelers, which pervade the present text.

The systems perspective

Systems, models, and modeling

Chapter outline

1.1 Why model ecological systems?

Why do we model ecological systems? Why do we model any system? Most often, we use simulation models, the focus of this book, to try to predict the future. Perhaps the most familiar everyday example of such use is to simulate weather patterns to try to answer the question "will it rain tomorrow?" But we also frequently use models to explore the possible cause-effect relationships within complex systems that generate system behavior. This exploration of the interactions among system components is a requisite step in the modeling process linked directly to our desire to predict specific aspects of system behavior. Where will hurricane X make landfall? Or we may investigate the effects on model behavior of a wide array of hypothetical relationships among system components with the goal of identifying plausible, as opposed to implausible, sets of interactions that could generate a certain type of system dynamics. What causal relationships could generate "the perfect storm?"

Two important points related to these uses of models are (1) it is impossible to predict the future with certainty, and (2) we often learn more about the system we are studying from the process of developing a model than from the final model itself. Strictly speaking, it is, even theoretically, impossible to predict the future states of a system unless we are dealing with a completely closed system, and such systems do not exist. On the one hand, this is purely a matter of semantics. On the other hand, arguably some of the most serious problems associated with the expectations and interpretations of model "predictions" have resulted from a failure to communicate clearly what we mean when way say "the model predicts that …" What we mean is that, as a result of the logical and numerical relationships that comprise the model, these are the outputs of the model. This meaning is implicit in our use of "model predictions" throughout this book. The extent to which these model outputs represent reality always remains uncertain. However, we can quantify the uncertainty associated with model outputs. This often helps us reduce the range of plausible alternative future states of the system we need to take into consideration. Furthermore, we can identify the likely sources of model uncertainty, thereby suggesting what new data from the real system would be most valuable to obtain to improve the usefulness of the model.

With regard to the second point, the usefulness of ecological simulation modeling often results as much from the process (conceptual model formulation, quantitative model specification, and model evaluation) as from the product (the final model and simulations of system dynamics). An appreciation of the heuristic aspect of the modeling process, as well as the development of modeling skills, is gained primarily through practice. However, we must maintain a keen awareness of exactly what we are doing (in practice), why we are doing it (in theory), and why it makes (common) sense. Without this awareness, we risk making silly mistakes; even experienced modelers are not immune from these pitfalls, which often come hidden under a thick covering of sophisticated quantitative techniques and associated jargon. An underlying theme of this book is the "oneness" of systems theory, modeling practice, and common sense.

1.2 Some terminology

Speaking of jargon, we already have used several terms that we should describe before proceeding. We present the following as working definitions/descriptions that we will adhere to throughout this book.

1.2.1 System

Like many words for which we have an intuitive understanding, "system" is difficult to define precisely. In relation to the physical or biological sciences, a system is an organized collection of interrelated physical components characterized by a boundary and functional unity. A system is a collection of "communicating" materials and processes that together perform some set of functions. A system is an interlocking complex of processes characterized by many reciprocal cause-effect pathways.

Clearly, any collection of interacting objects can be viewed as a system. The principal attribute of a system is that we can understand it only by viewing it as a whole. Another important attribute of a system, or more strictly speaking, of our conceptualization of a particular system, is that it is chosen for a particular purpose: to answer a question, to demonstrate a theory, or to classify part of the natural world. We will refer often throughout this book to our "system-of-interest." Common examples of ecological systems include ecosystems, communities, populations, and individuals. But the rumen of a deer may be viewed as a system, as can the planet earth.

1.2.2 Systems perspective

The systems perspective, or "systems approach," embodies both a philosophical perspective and a collection of quantitative techniques that emphasize a holistic approach to solving problems involving complex systems. The systems perspective follows directly from our definition of a system in that it provides a view of systems as intact entities whose dynamics cannot be understood by examining their parts one by one in isolation from the rest of the system. For example, one cannot understand how a clock works by dismantling it and looking carefully at each individual part. This is not to say that the systems way of thinking about things is better than the reductionist way of thinking about things. The systems view is complimentary to the reductionist view. It provides a different lens through which to gain insight into world in which we live. Some things are better viewed through the lens of a microscope, some through the lens of a telescope, some through the systems lens.

1.2.3 Complex system

As with the concept represented by "system," we have an intuitive understanding of the concept represented by "complex system," but the concept is difficult to define precisely. There is a relatively recent and rapidly growing field called complexity science that attests to this difficulty. However, for our purposes, a good working definition is simply that a complex system is one whose behavior is impossible to understand intuitively or via our brain's unaided reasoning process. That is, we cannot understand system behavior by simply observing it and thinking about it. This inability to understand results more from indirect effects caused by the interconnectedness of system components than from the sheer number of system components.

1.2.4 Model

A model is an abstraction of reality. It is a formal description of the essential elements of a problem, a formal description of the system in which we are interested. The description can be physical, mathematical, or verbal, although some modelers object to the idea of "word models" because language can be so ambiguous. Mathematical models can be classified in a variety of ways. Some of the more relevant dichotomies for our purposes include (1) dynamic versus static, (2) deterministic versus stochastic, (3) correlative versus explanatory, and (4) simulation versus analytical.

Dynamic versus static

Models may or may not represent systems that change with time. A static model describes a relationship or set of relationships that do not change with time. Common examples include regression models and species distribution models. A dynamic model describes a time-varying relationship. Examples include simulation models, which we will discuss further in a moment.

Deterministic versus stochastic

A model is deterministic if it contains no random variables. Deterministic model predictions under a specific set of conditions are always exactly the same. A model is stochastic if it contains one or more random variables. Stochastic model predictions under a specified set of conditions are not always exactly the same, because random variables within the model potentially can take on different values each time the model is solved. The values of the random variables might be drawn from statistical distributions (e.g., uniform, normal, binomial) or from cumulative frequency distributions (e.g., based on historical data).

The choice between using a deterministic versus a stochastic model depends on the specific objectives of the modeling project. Deterministic models generally are easier to build because they require only single point estimates of values of model components. Stochastic models require specification of complete distributions from which to randomly draw values of model components. Deterministic models also are easier to use because predictions for a given situation need be made only once. When using stochastic models, predictions for a given situation must be replicated to obtain an adequate sample of the range of responses for the given situation. However, if project objectives require explicit representation of variability, either the variability associated with the estimates of system parameters or the inherent variability of the system itself, we must use a stochastic model. Likewise, if we wish to make statistical comparisons of model predictions for different situations, we must use a stochastic model.

Correlative versus explanatory

Correlative models are developed primarily to describe and summarize a set of relationships, without regard for the appropriate representation of processes or mechanisms that operate in the real system. The goal is prediction, not explanation. A model predicting metabolic rate of an animal solely as a function of body mass is an example. Explanatory models are developed primarily to represent the internal dynamics of the system-of-interest appropriately. The goal is explanation through representation of the causal mechanisms underlying system behavior. A model representing the metabolic rate of an animal as a function of body mass, level of activity, environmental temperature, wind, and length of exposure to ambient conditions is an example. Solely explanatory models and solely correlative models form two ends of a continuum, and classification relative to this dichotomy is based more on the intent of the modeler than on the structure of the model. A model that we view as explanatory at one level of detail we might view as correlative at a finer level of detail. A model representing annual population recruitment as a function of population size appears explanatory compared to a model representing annual recruitment simply as a constant determined by averaging historical data. But this same model appears correlative compared to a model that calculates recruitment based on age-specific natality rates of individuals within the population, which, in turn, are based on the individuals' social rank and nutritional status during the breeding season.

Simulation versus analytical

Models that can be solved in closed form mathematically are analytical models. Some models composed of linear differential equations fall into this category. For such models, a general solution can be obtained that applies to all situations the model can represent. The model of exponential population growth in an unlimited environment seen in most basic ecology texts is an example ($N_t = N_0 e^{rt}$, where N_t is population size at time t and r is the instantaneous rate of population increase). Assuming we know the initial population size (N_0) and the instantaneous rate of increase (r), we can solve this model for population size at any time (N_t) by substituting the desired value for t into the model. For example, if the initial population size is 100 individuals and the intrinsic rate of increase is 0.1, we can calculate the population size at time 5 ($100e^{(0.1 \times 5)} = 164.9$) without the need to calculate the intervening population sizes.

Models that have no general analytical solution must be solved numerically for each successive time step using the set of arithmetic operations specified in the model. Simulation models fall into this category and many ecological models are of this type. For example, if the instantaneous rate of population increase in the previous model was represented as some function of current population size and environmental conditions ($r = f(N_t, E_t)$) which was too complex to solve analytically, to calculate population size at time 5 we would need to make successive numerical calculations for population size at time 1, time 2, time 3, and time 4. That is, we would need to calculate the change in the size of the population over some finite time interval based on population size at the beginning of that time interval, and then update population size and advance to the beginning of the next time interval. We will take a more detailed look at simulation, including a numerical example, in the next section.

Philosophically, the choice between analytical models and simulation models involves deciding whether we sacrifice ecological realism to obtain an analytical model or sacrifice mathematical power to include more ecological realism. From a more practical standpoint, these considerations are no less interesting and are influenced primarily by objectives of the modeling project. If the level of detail at which the system-of-interest must be represented to meet our objectives permits use of an analytical model, then we should use an analytical model. However, if the appropriate level of detail requires a model too complex to be represented in an analytically solvable form, then we should use a simulation model. In most cases, in ecology and natural resource management, we will need to represent the system-of-interest in a mathematical manner too complex for analytical treatment.

1.2.5 Simulation

Simulation is the process of using a model to mimic, or trace through step by step, the behavior of the system-of-interest. Simulation models are composed of a series of arithmetic, and perhaps logical, operations that together represent the structure (state) and behavior (change of state) of the system-of-interest. These two very simple yet powerful concepts of the state of the system and the change of state of the system are fundamental to simulation. The system-of-interest exists in different states at different points in time and rules govern the manner in which the state of the system changes as time passes. The rules governing change also may change from time to time because they are themselves a function of the state of the system. If we choose appropriate variables to describe the system and appropriately represent the rules governing change, then we should be able to trace the state of the system through time, which is to say we can simulate behavior of the system.

As a simple numerical example of simulation, suppose we wanted to simulate changes in weight of an animal over a 2-day period, assuming that rate of food consumption was a function of animal weight and that respiration rate was a function of animal weight and environmental temperature. (We will assume that digestive and assimilation efficiencies are 100% and that respiration is represented in biomass equivalents of energy respired.) We might describe the current state of our system-of-interest in terms of the current weight of the animal (W_t, in g) and the change in state of our system in terms of the amount of food consumed (*con*, in g day^{-1}) and the biomass equivalent of the amount of energy respired (*res*, in g day^{-1}) at a given

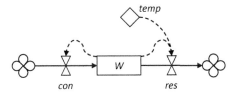

FIG. 1.1

Diagram of a simulation model predicting changes in weight (W) of an animal over time. Rate of food consumption (*con*) is affected by animal weight, and respiration rate (*res*) is affected by animal weight and environmental temperature (*temp*).

environmental temperature ($temp_t$, in °C) (Fig. 1.1). The symbols used in Fig. 1.1 have specific meanings that we will define formally in Chapter 2. For now, note only that the figure is a simple diagrammatic representation of the weight fluctuation model just described. We might appropriately represent the rules governing changes in the state of the system using two equations, one representing *con* as a function of W_t and one representing *res* as a function of W_t and $temp_t$.

$$W_{t+1} = W_t + (con - res)\,\Delta t \qquad (1.1)$$

$$con = f(W_t) \qquad (1.2)$$

$$res = f(W_t, temp_t) \qquad (1.3)$$

These are difference equations, and the Δt in Eq. (1.1) represents 1 finite time unit, 1 day in this example. To quantify the functional relationships represented in Eqs. (1.2) and (1.3), let us suppose that

$$con = 0.05 \times W_t \qquad (1.4)$$

and

$$res = 0.0025 \times temp_t \times W_t \qquad (1.5)$$

where $W_0 = 100$, $temp_0 = 20$, and $temp_1 = 19$. Note that $temp_0 = 20$ means that temperature from the beginning of time period 0 to the beginning of time period 1, the first day we are going to simulate, is 20, and $temp_1 = 19$ means that temperature from the beginning of time period 1 to the beginning of time period 2, the second day we are going to simulate, is 20. Thus, for the first day,

$$con = 0.05 \times 100 = 5 \qquad (1.6)$$

$$res = 0.0025 \times 20 \times 100 = 5 \qquad (1.7)$$

$$W_1 = 100 + (5 - 5) \times 1 = 100 \qquad (1.8)$$

and for the second day,

$$con = 0.05 \times 100 = 5 \tag{1.9}$$

$$res = 0.0025 \times 19 \times 100 = 4.75 \tag{1.10}$$

$$W_2 = 100 + (5 - 4.75) \times 1 = 100.25 \tag{1.11}$$

If we wanted to simulate changes in weight of the animal during a third day, and then a fourth day, first, we would need to specify the environmental temperatures during each of those days, then we would repeat the same sequence of calculations that we performed for the first 2 days we simulated. If we wanted to simulate changes in weight over a 365-day period, the procedure would be the same, although we probably would want to instruct a computer how to do these calculations.

1.3 Simulation modeling, laboratory experiments, and field observations

Within the context of scientific research, how is simulation modeling related to laboratory experimentation and field observations? Just as laboratory experiments compliment field observations, and vice versa, simulation modeling provides a complimentary perspective on the system we are studying which would be difficult, if not impossible, to obtain in the laboratory or in the field. The difficulty often arises because the data required to test our hypotheses would be too costly, would take too long, or would be too dangerous to collect in the real system. In such cases, we can conduct simulated experiments with a model of the system-of-interest in a manner analogous to the manner in which we conduct experiments in the laboratory or make observations in the field (Fig. 1.2). Just as we focus on, or "abstract," particular aspects of the real world to permit more detailed studies in the laboratory or the field, we abstract particular aspects of the real world to form a simulation model. We develop an experimental design for simulations in the same manner that we develop experimental designs for laboratory or field studies. Likewise, we analyze simulation results, simulated data, using the same qualitative, quantitative, and/or statistical methods that we use to analyze data collected in the laboratory or in the field. And, of course, we subject simulation results to the same scrutiny to which we subject laboratory and field results with regard to biases and uncertainties associated with our data.

1.4 Basic characteristics of systems

Systems, ecological or otherwise, all have five important characteristics in common: (1) A system is more than the sum of its parts. (2) System structure is the source of system behavior. (3) Systems are hierarchical. (4) Systems are resilient. (5) Systems are self-organizing.

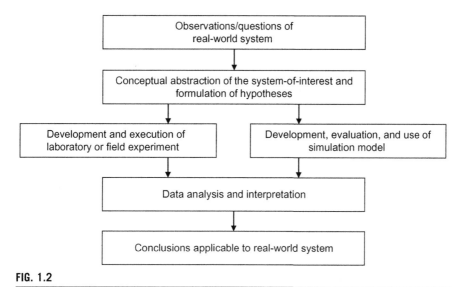

FIG. 1.2

Simulation modeling related to laboratory experiments and field observations.

Modified from Van Dyne, G.M., 1969. Grasslands Management, Research, and Training Viewed in a Systems Context. Colorado State University, Fort Collins, CO.

1.4.1 More than the sum of its parts

A system is more than the sum of its parts in that we can understand its behavior only by viewing it as a whole. This is the principal attribute of a system and invariably is a part of its definition for our purposes. We could think of two competing soccer teams as a system with the system components being the ball, the field, the players, and the coaches, and the interactions among these components involving the laws of physics, the communication among the players, and the strategies of the coaches. If we observed the behavior of any given player in isolation from the others during the course of a match, it would be virtually impossible to understand his/her behaviors. However, if we observed these behaviors within the context of the behaviors of all of the other players, assuming we knew the objective and the rules of the game, the behaviors would be much more understandable.

1.4.2 Structure is the source of behavior

That system structure is the source of system behavior may not be intuitively obvious. We are cautioned, and rightfully so, that inferring particular processes from particular patterns, and vice versa, is risky business. Nonetheless, the general behavior of systems is inherent in their structure. An elegantly simple classroom demonstration of the behavior inherent in system structure has been provided by Meadows (2008) using a Slinky. (For those who do not remember when there was no Internet, a Slinky is a toy consisting of a long, loose spring that can be made to bounce up and

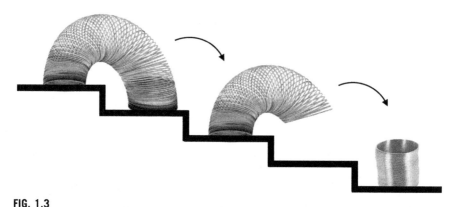

FIG. 1.3

A Slinky "walking" itself downstairs.

down, or pour back and forth from one hand to the other, or walk itself downstairs (Fig. 1.3).) The demonstration consisted of placing the Slinky on one upturned palm, grasping the top coil of the Slinky with the other hand, and then pulling the bottom hand away. The lower end of the Slinky would drop, bounce back up again, and then "yo-yo" up and down, suspended by the top hand. Professor Meadows would ask her students to explain what made the Slinky bounce up and down. They would respond "you took your hand away." She then would repeat the experiment with the box the Slinky came in, placing the box on one upturned palm, grasping the top of the box with the other hand, and pulling the bottom hand away. Of course, nothing would happen. Why did nothing happen to the box and what was the source of the behavior of the Slinky? The obvious (at least after the fact) answer was that the structure of the box was different than the structure of the Slinky and source of the behavior of the Slinky was its own structure.

1.4.3 **Hierarchical**

Systems are hierarchical in that each system is a functional component of a larger system and is composed of functional components consisting of smaller systems. All systems are connected directly or indirectly, thus system boundaries are arbitrary and permeable. The hierarchical relatedness of systems that encompass vastly different spatial areas and whose behavior unfolds over vastly different time scales are illustrated impressively by a short film, *the Powers of Ten* (http://www. eamesoffice.com/the-work/powers-of-ten/). The film begins with the camera focused on a 10 m by 10 m area viewed from a height of 10 m. From this perspective, we see a man and a woman seated outdoors on a beach blanket with a picnic basket. As the film rolls forward the perspective changes from 10 m by 10 m from a height of 10 m, to 100 m by 100 m from a height of 100 m, to 1000 m by 1000 m from a height of

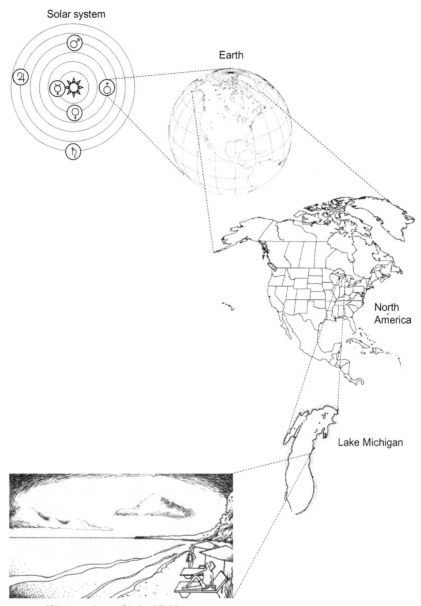

Solar system

Earth

North America

Lake Michigan

Western shore of Lake Michigan

FIG. 1.4

The hierarchical relatedness of systems that encompass vastly different spatial areas and whose behavior unfolds over vastly different time scales as illustrated by the film titled the "Powers of Ten" (The drawing of "Western shore of Lake Michigan" by Rui-Pu Wang).

1000 m, and so on (Fig. 1.4). We see that the man and the woman are on a beach next to a large lake near a big city. As the contours of the lake come into view, we see that they are on the western shore of Lake Michigan, near Chicago or Milwaukee. The contours of North America then become recognizable, and then those of the Western Hemisphere. The entire planet Earth comes into view and fades out of sight, as does the sun. The stars in the Milky Way come into view and fade as the galaxy becomes but a tiny spot of light. The film then brings us flying back to the original perspective and begins the reverse journey, changing the perspective from 10 m by 10 m from a height of 10 to 1 m by 1 m from a height of 1 m, and so on. It turns out the camera is aimed at the back of the man's hand. We see the hairs on his hand, we pass through various layers of skin, and finally we find ourselves in the nucleus of a carbon atom in the center of the hand. There the film ends.

We (the authors) always ask students in our ecological modeling class to identify the most "real" view of reality seen in the film. Where should we stop the film if we want to view the real system most appropriately? What is the best perspective? Most students provide a specific answer, often recognizable as a traditional ecosystem-, community-, or population-level perspective. Others simply return an incredulous stare (perhaps wondering about the mental health of their instructors). However, after we mention that we are most interested in the dispersion pattern of the people surrounding the couple on the beach, everyone agrees where we should stop the film. Perhaps not on the exact frame we should be looking at, but no one suggests we should be looking at the stars in the Milky Way or at the hairs on the back of the man's hand. Thus, the obvious (at least after making explicit our main interest) correct answer was "it depends on what you are interested in."

1.4.4 **Resilient**

Resilience has many definitions depending on the branch of engineering, ecology, or systems science doing the defining. In everyday language, resilience is the ability to bounce back. A spring is resilient because it regains its original shape after being stretched. The opposite of resilience is rigidity or brittleness. For our purposes, resilience is the ability of a system to survive and persist in a recognizable form within a variable environment. Resilience is not the same as being static or constant over time, and the distinction between static stability and resilience is important. Resilient systems most often are dynamic, exhibiting short- and long-term oscillations, as well as irregular episodes of collapse and recovery in response to external perturbations. System resilience is difficult to observe because it only can be identified from a perspective that encompasses the entire spectrum of system behaviors. It is easier to identify the lack of resilience, which manifests itself when a system collapses and fails to recover. There always are limits to system resilience.

A simple physical example of resilience and the limits thereof is illustrated in Fig. 1.5. Dotted arrows on the uneven surface of hills and valleys represent possible states of a system, whose current state is represented by the ball. In Fig. 1.5A, the ball

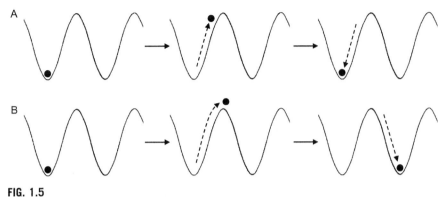

FIG. 1.5

A simple physical example of resilience and the limits thereof. Dotted arrows on the uneven surface of hills and valleys represent possible states of a system, whose current state in each diagram is represented by the ball.

is pushed halfway up the hill by an external force, is maintained in that position until the force is removed, and then rolls back down the hill to its original position. In Fig. 1.5B, the ball is pushed to the top of the hill by an external force, and then just past the top of the hill, after which it rolls downhill into the adjacent valley where it remains. To complete analogy, the ball no longer is recognizable from the perspective at the bottom of the first valley.

Resilience arises in complex systems from a rich structure of feedback loops. We will take a close look at the operation of feedback loops in Chapter 2, but for now, a simple example suffices. An initially small population growing in a limited environment exhibits the results of both a positive feedback loop and a negative feedback loop (Fig. 1.6). While the population is relatively small and resources are abundant, the larger the population gets, the more rapidly it grows. A positive, self-reinforcing, feedback loop predominates, and the population increases at an increasing rate. As the population becomes larger and larger and resources become scarce, it reaches a point beyond which the larger the population gets, the more slowly it grows. A negative, self-limiting, feedback loop predominates, and the population increases at a decreasing rate. (Note that negative feedback does not imply the entity being affected is decreasing, but rather that the rate at which the entity is changing is decreasing.)

1.4.5 Self-organizing

Self-organization is the capacity of a system to reorganize its own structure, most often making its own structure more complex. Self-organization is so ubiquitous that we tend to take it for granted. A seed self-organizes into a plant. An egg self-organizes into a chicken. A group of wolves self-organizes into a wolf pack. A group of species self-organizes into an ecological community. Such examples perhaps imply that the rules of self-organization producing complex systems necessarily

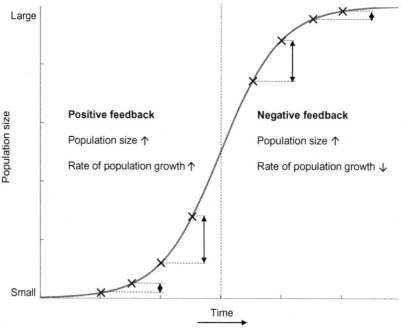

Positive feedback

Population size ↑

Rate of population growth ↑

Negative feedback

Population size ↑

Rate of population growth ↓

FIG. 1.6

An initially small population growing in a limited environment exhibiting the results of both a positive feedback loop and a negative feedback loop. At smaller population sizes, a positive, self-reinforcing, feedback loop predominates, and the population increases at an increasing rate. As the population becomes larger, a negative, self-limiting, feedback loop predominates, and the population increases at a decreasing rate.

are complex. But the formation of a snowflake also is an example of self-organization giving rise to a complex structure, as is the formation of complex chemical compounds resulting from a single original autocatalytic set. In fact, recent advances in complexity science confirm that complex systems can arise as the result of quite simple rules of self-organization. Science, which, by the way, is itself a self-organizing system, currently hypothesizes that all complexity arises from simple rules. This remains to be seen.

Regardless of its genesis, self-organization produces heterogeneity in system structure and unpredictability in system behavior. In the opening section of this chapter, we cautioned that it is impossible to predict the future with certainty. We added that even theoretically, it is impossible to predict the future states of a system unless it is a completely closed system, and that such systems do not exist. Self-organization guarantees the lack of closure of system structure, and the fact that system structure is the source of system behavior thus guarantees the lack of predictability of system behavior.

1.4.6 **Characteristics of systems models**

This is not to say that all of our models of systems, all of the systems-of-interest that we define to address our questions, necessarily will exhibit each of the five system characteristics described above. By definition, all will exhibit the first two. But many simple models, such as the majority of those we use in this book, as well as many useful ecological models found in the literature, are not explicitly hierarchical, self-organizing, or resilient. This is simply because the particular questions these models were developed to address did require consideration of these system characteristics. Thus, such models were bounded arbitrarily in space and time and with regard to the detail with which system components were represented such that these characteristics could not manifest themselves. Often, this is appropriate. However, we should keep in mind that conceptually all ecological systems are implicitly hierarchical and latently self-organizing and resilient.

1.5 **Sources of surprising systems behavior**

An understanding of the basic characteristics of systems presented in the previous section does not prevent us from being surprised by the behavior of even relatively simple systems. This is because our minds can keep track of only a few variables at a time. Our mental models of how we think the world works are based on what we can see, hear, touch, smell, taste, deduce, and intuit. They usually have a strong congruence with the real world, but they fall short of representing the real world completely. Thus, we often draw illogical conclusions from accurate assumptions or logical conclusions from inaccurate assumptions. Meadows (2008) suggested several reasons why systems behavior is so often surprising. These reasons perhaps have as much to do with how our mental models fail to take into account the complications of the real world as they do with the characteristics of systems per se. Surprises come from failure to be aware of (1) nonlinear relationships, (2) nonexistent system boundaries, (3) layers of limiting factors, (4) ubiquitous delays between cause and effect, and (5) bounded rationality.

1.5.1 **Nonlinear relationships**

Linear relationships are easy to think about. A linear relationship between two components in a system can be drawn on a graph as a straight-line. The slope of the line is a constant, that is, it is a relationship with constant proportions. Sets of linear equations often are solvable analytically. Linear systems can be taken apart and put back together, the parts add up. A nonlinear relationship is one in which the change in one component does not produce a proportional effect on the other component. The relationship cannot be drawn on a graph with a straight line. The line will have curves and may sometimes go up and sometimes go down. Sets of nonlinear equations are difficult to solve and may be unsolvable analytically. These nonlinearities surprise us because they foil the reasonable expectation that if a little of something is good, then a lot of that same thing will be better. If last year a kilogram of fertilizer was good for the plants in my garden, then this year a ton of fertilizer will be even better. Or if a little destructive action caused only a tolerable amount of harm, then a little more of that same action will cause only a little more harm. If last year I dumped a small amount of leftover fertilizer in my pond and there was only a small algal bloom, then this year I can dump a slightly larger amount of leftover fertilizer in my pond and there will be only a slightly larger algal bloom. We tend to be linear minds living in a nonlinear world. We need to train ourselves to look for and expect to find non-linear relationships in the systems we study and in which we live.

1.5.2 Nonexistent system boundaries

Implicit in our use of the popular term "side-effects" is "effects that we had not fore-seen or do not want to think about." Systems rarely have real boundaries. Everything is connected to everything else, often indirectly, and often not neatly. Yet necessarily models of our system-of-interest must have boundaries. If we are to understand any-thing, we have to simplify. We have to invent boundaries for clarity, and for sanity. However, boundaries can create problems when we forget that we have created them artificially. If we draw boundaries too narrowly, systems surprise us. For example, if a model we develop to aid planning for a national park did not consider anything outside of the physical park boundary, we would be surprised by the effects of rivers, of migrating wildlife, and perhaps of nomadic peoples which enter and leave the park. On the other hand, if we draw boundaries too broadly, we will become hope-lessly lost in irrelevant details. For example, in our park planning model, we would be ill-advised to include a detailed representation of the earth's climate to explore seasonal weather effects on park visitation. Unfortunately, we often lack the mental flexibility to custom fit system boundaries to the problem at hand. We are attached to customary boundaries, which often are associated with academic disciplines. We must strive to find the appropriate system boundaries for thinking about each new problem.

1.5.3 Layers of limiting factors

We like to think about single causes neatly producing single effects. But we live in a world in which many causes routinely converge to produce many effects. Multiple inputs produce multiple outputs and virtually all of the inputs, and therefore, virtually all of the outputs, are limited. Agronomists and ecologists, among others, are familiar with Liebig's "law of the minimum," which states that growth is limited not by total resources available but by the scarcest resource available (the limiting factor). But the concept of a limiting factor is widely misunderstood. Although adding nitrogen will not increase growth of a plant that lacks phosphorus, adding phosphorus might only serve to make micronutrient "x" the most limiting. Perhaps the addition of phos-phorus would cause an increase in soil microbes that also utilize nutrient "x" thus depleting its supply. Subsequent addition of nutrient "x" might shift the species composition of the microbial community in favor of a species harmful to the plant, which thus becomes limiting to plant growth. There are layers of limits surrounding any growing entity with multiple inputs. Insight comes not from identifying the current limiting factor, but from appreciating, that growth itself depletes or enhances supplies of the various potentially limiting factors. That is, growth itself changes what is limiting, and there always are limits to growth. In coupled human-natural systems, if man does not impose the limits, the limits will be imposed by the system.

1.5.4 Ubiquitous delays between cause and effect

We learn at an early age that cause and effect are tightly linked in time and space. We touch a flame, and we are burned immediately. This rule works well in very simple systems. But delays between cause and effect are ubiquitous in more complex systems. We often are surprised by how long it takes to accomplish a simple task due to hidden delays. We often are surprised when we suddenly become ill several days after brief exposure to a colleague with the flu. Delays influence system behavior in ways that cannot be ignored. Oscillations, overshoots, and collapses always are caused by delays. Predator-prey population oscillations and overshoot-collapse cycles of large herbivores and their forage resources are classic examples often mentioned in introductory ecology texts. Changing the length of a delay can change system behavior in dramatic ways. The length of delay between arrival of a crop pest and the initiation of pest control can determine the order of magnitude of crop damage. Just as identification of appropriate boundaries for thinking about a system depends on the problem at hand, so does identification of important delays. If we are concerned about oscillations in system behavior that take weeks, we probably do not need to represent delays that take minutes, or years. Thus, the identification and representation of delays at appropriate temporal scales is crucial if we are to construct useful models of complex systems.

1.5.5 Bounded rationality

Bounded rationality means that we make reasonable decisions based on the information we have, not on the information we do not have. Although this is to state the obvious, it has implications that we commonly ignore. Tourists flock to remote natural areas and complain that those places have been ruined by tourists. Fishermen overfish and destroy their own livelihood. Traveling to remote natural areas to enjoy the solitude and catching more fish to sell are reasonable decisions based on available information. But we do not have complete or perfect information, especially about more distant parts of the system. Fishermen do not know how many fish there are, much less how many fish other fishermen will catch. Ecologists will recognize the latter as a classic example of the "tragedy of the commons." Although we usually think of bounded rationality in terms of human behavior, the idea is applicable in a figurative sense to other system components. Information about what is happening in a given part of the system is not universally available to all other system components. Migratory birds do not know what conditions await them on their breeding grounds. Furthermore, information may be delayed and/or distorted between its points of origin (cause) and ultimate use (effect). In the predator-prey and large herbivore examples in the previous section, reproduction rates of both the predators and large herbivores resulted from "misinterpretation" of information about the size of the available resource base. That is, the per capita availability of prey and forage, respectively, changed between reproduction (conception) and recruitment (birth). Thus, appreciation of the importance of appropriate representation of the "information network" within systems is a requisite for the construction of useful models of complex systems.

Basic building blocks of system structure and behavior

2

Chapter outline

In the previous chapter, we noted that one of the basic characteristics of systems is that system structure is the source of system behavior. In this chapter, we focus on the practical mechanics of representing system structure, and on how system behavior is related to system structure. We present several modules composed of simple

Developments in Environmental Modeling, Volume 31, ISSN 0167-8892, https://doi.org/10.1016/B978-0-444-64163-2.00002-5

combinations of basic system components which provide useful building blocks for constructing a wide variety of system structures. System behaviors generated by these simple modules include (1) linear growth and decline, (2) exponential growth and decline, (3) logistic growth, (4) overgrowth and collapse, (5) oscillations, and (6) time lags. The intent is not to be exhaustive, but rather to present a relatively few basic building blocks that in our experience have proven useful as a point of departure for modeling many systems.

2.1 Representation of the building blocks

We first briefly define a set of symbols that we will use to represent the building blocks of system structure diagrammatically. We will consider the meanings and use of these symbols in more detail in Part 2 of this book when we discuss the modeling process more formally. But for now, the following definitions will suffice for representation of the building blocks. (1) *State variables:* State variables represent points of accumulation of material within the system. (2) *Material transfers:* A material transfer represents the movement of material from outside the system into a state variable, between two state variables, or from a state variable out of the system. (3) *Sources and sinks:* Sources and sinks represent origination and termination points, respectively, of material transfers into and out of the system. (4) *Information transfers:* Information transfers represent the use of information about the state of the system to control the change of state of the system. (5) *Driving variables:* Driving variables affect, but are not affected by, the rest of the system. (6) *Constants:* Constants are numerical values describing important characteristics of a system that do not change, or that can be represented as unchanging. (7) *Auxiliary variables:* Auxiliary variables most commonly represent processes or concepts in the system-of-interest that we wish to indicate explicitly, which otherwise would be implicit in the information transfers among model components (constants, driving variables, state variables, material transfers).

In the following sections, we present the building blocks categorized by structure (number of state variables and material transfers) and by behavioral characteristics (growth, decline, oscillations, and positive or negative feedback). Positive feedback usually is thought of as referring to a situation in which the rate of material transfer becomes faster as the state variable becomes larger, and negative feedback usually is thought of as referring to a situation in which the rate of material transfer becomes slower as the state variable becomes larger. However, note that positive feedback also can refer to a situation in which the rate of material transfer becomes slower as the state variable becomes smaller, and negative feedback can refer to a situation in which the rate of material transfer becomes faster as the state variable becomes smaller. Diagrams and equations for each of the following modules are presented in Fig. 2.1.

A

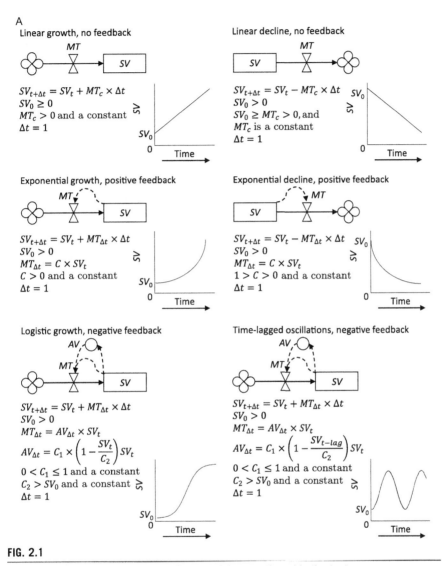

FIG. 2.1

Diagrams and equations for each of the modules mentioned in the text.

(Note that we will use a compartment model format to represent these modules, as well as the models we present throughout this book. We have chosen this "box-and-arrow" format because it is widely used and, in our experience, the easiest to understand. However, there are several other commonly used ecological modeling formats, including matrix models, network models, and individual-/agent-based models, which we will describe and compare in Chapter 14.)

B Linear growth or decline, no feedback

$SV_{t+\Delta t} = SV_t + \left(MT_{in_{\Delta t}} - MT_{out_{\Delta t}}\right) \times \Delta t$
$SV_0 > 0$
$MT_{in_{\Delta t}} > 0$ and a constant
$MT_{out_{\Delta t}} > 0$ and a constant
$\Delta t = 1$

Exponential growth or decline, positive feedback

$SV_{t+\Delta t} = SV_t + \left(MT_{in_{\Delta t}} - MT_{out_{\Delta t}}\right) \times \Delta t$
$SV_0 > 0$
$MT_{in_{\Delta t}} = C_1 \times SV_t$
$MT_{out_{\Delta t}} = C_2 \times SV_t$
$C_1 > 0$ and a constant
$C_2 > 0$ and a constant
$\Delta t = 1$

Logistic growth, negative feedback

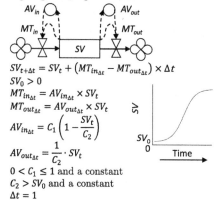

$SV_{t+\Delta t} = SV_t + \left(MT_{in_{\Delta t}} - MT_{out_{\Delta t}}\right) \times \Delta t$
$SV_0 > 0$
$MT_{in_{\Delta t}} = AV_{in_{\Delta t}} \times SV_t$
$MT_{out_{\Delta t}} = AV_{out_{\Delta t}} \times SV_t$
$AV_{in_{\Delta t}} = C_1 \left(1 - \dfrac{SV_t}{C_2}\right)$
$AV_{out_{\Delta t}} = \dfrac{1}{C_2} \cdot SV_t$
$0 < C_1 \le 1$ and a constant
$C_2 > SV_0$ and a constant
$\Delta t = 1$

C Renewable state variable constrained by a nonrenewable
state variable, unsustainable growth

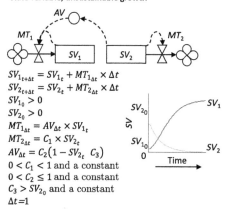

$SV_{1_{t+\Delta t}} = SV_{1_t} + MT_{1_{\Delta t}} \times \Delta t$
$SV_{2_{t+\Delta t}} = SV_{2_t} + MT_{2_{\Delta t}} \times \Delta t$
$SV_{1_0} > 0$
$SV_{2_0} > 0$
$MT_{1_{\Delta t}} = AV_{\Delta t} \times SV_{1_t}$
$MT_{2_{\Delta t}} = C_1 \times SV_{2_t}$
$AV_{\Delta t} = C_2 \left(1 - SV_{2_t} \; C_3\right)$
$0 < C_1 < 1$ and a constant
$0 < C_2 \le 1$ and a constant
$C_3 > SV_{2_0}$ and a constant
$\Delta t = 1$

FIG. 2.1—CONT'D

D Renewable state variable constrained by a nonrenewable state variable, unsustainable growth and collapse

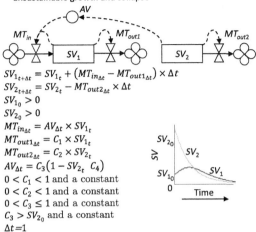

$$SV_{1_{t+\Delta t}} = SV_{1_t} + \left(MT_{in_{\Delta t}} - MT_{out1_{\Delta t}}\right) \times \Delta t$$
$$SV_{2_{t+\Delta t}} = SV_{2_t} - MT_{out2_{\Delta t}} \times \Delta t$$
$$SV_{1_0} > 0$$
$$SV_{2_0} > 0$$
$$MT_{in_{\Delta t}} = AV_{\Delta t} \times SV_{1_t}$$
$$MT_{out1_{\Delta t}} = C_1 \times SV_{1_t}$$
$$MT_{out2_{\Delta t}} = C_2 \times SV_{2_t}$$
$$AV_{\Delta t} = C_3\left(1 - SV_{2_t}\ C_4\right)$$
$0 < C_1 < 1$ and a constant
$0 < C_2 < 1$ and a constant
$0 < C_3 \leq 1$ and a constant
$C_3 > SV_{2_0}$ and a constant
$\Delta t = 1$

E   Renewable state variable constrained by a renewable state variable, oscillations

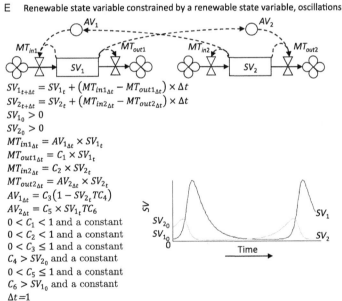

$$SV_{1_{t+\Delta t}} = SV_{1_t} + \left(MT_{in1_{\Delta t}} - MT_{out1_{\Delta t}}\right) \times \Delta t$$
$$SV_{2_{t+\Delta t}} = SV_{2_t} + \left(MT_{in2_{\Delta t}} - MT_{out2_{\Delta t}}\right) \times \Delta t$$
$$SV_{1_0} > 0$$
$$SV_{2_0} > 0$$
$$MT_{in1_{\Delta t}} = AV_{1_{\Delta t}} \times SV_{1_t}$$
$$MT_{out1_{\Delta t}} = C_1 \times SV_{1_t}$$
$$MT_{in2_{\Delta t}} = C_2 \times SV_{2_t}$$
$$MT_{out2_{\Delta t}} = AV_{2_{\Delta t}} \times SV_{2_t}$$
$$AV_{1_{\Delta t}} = C_3\left(1 - SV_{2_t} TC_4\right)$$
$$AV_{2_{\Delta t}} = C_5 \times SV_{1_t} TC_6$$
$0 < C_1 < 1$ and a constant
$0 < C_2 < 1$ and a constant
$0 < C_3 \leq 1$ and a constant
$C_4 > SV_{2_0}$ and a constant
$0 < C_5 \leq 1$ and a constant
$C_6 > SV_{1_0}$ and a constant
$\Delta t = 1$

FIG. 2.1—CONT'D

F "Main chain" of state variables, time-lagged oscillations

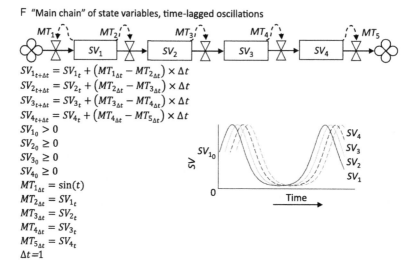

$$SV_{1_{t+\Delta t}} = SV_{1_t} + \left(MT_{1_{\Delta t}} - MT_{2_{\Delta t}}\right) \times \Delta t$$
$$SV_{2_{t+\Delta t}} = SV_{2_t} + \left(MT_{2_{\Delta t}} - MT_{3_{\Delta t}}\right) \times \Delta t$$
$$SV_{3_{t+\Delta t}} = SV_{3_t} + \left(MT_{3_{\Delta t}} - MT_{4_{\Delta t}}\right) \times \Delta t$$
$$SV_{4_{t+\Delta t}} = SV_{4_t} + \left(MT_{4_{\Delta t}} - MT_{5_{\Delta t}}\right) \times \Delta t$$
$$SV_{1_0} > 0$$
$$SV_{2_0} \geq 0$$
$$SV_{3_0} \geq 0$$
$$SV_{4_0} \geq 0$$
$$MT_{1_{\Delta t}} = \sin(t)$$
$$MT_{2_{\Delta t}} = SV_{1_t}$$
$$MT_{3_{\Delta t}} = SV_{2_t}$$
$$MT_{4_{\Delta t}} = SV_{3_t}$$
$$MT_{5_{\Delta t}} = SV_{4_t}$$
$$\Delta t = 1$$

FIG. 2.1—CONT'D

2.2 One state variable, one material transfer

2.2.1 Linear growth, no feedback

This module consists of one state variable and one material transfer that enters the state variable from a source (Fig. 2.1A). The rate of material transfer is constant and, thus, does not depend on the level of accumulation in the state variable. The state variable increases at a constant rate forever.

2.2.2 Linear decline, no feedback

This module consists of one state variable and one material transfer that leaves the state variable to a sink (Fig. 2.1A). As before, the rate of material transfer is constant and, thus, does not depend on the level of accumulation in the state variable. The state variable decreases linearly at a constant rate forever, obviously resulting in negative values for the state variable at some point in time.

2.2.3 Exponential growth, positive feedback

This module is the same as the linear increase module except that the rate of material transfer is a constant proportion of the level of the state variable (Fig. 2.1A). This results in a positive feedback loop and both the level of the state variable and the rate of material transfer increase exponentially forever.

2.2.4 **Exponential decline, positive feedback**

This module is the same as the linear decrease module except that the rate of material transfer is a constant proportion of the level of the state variable (Fig. 2.1A). This results in a positive feedback loop and both level of the state variable and the rate of material transfer decrease exponentially forever. Note that the term positive feedback refers to the nature of the change, that is, increasing at an increasing rate or decreasing at a decreasing rate, not to the direction of the change.

2.2.5 **Logistic growth, negative feedback**

There are a variety of modules that produce logistic (or sigmoid) growth. The common characteristic of these modules is the negative feedback loop from the state variable to rate of growth (Fig. 2.1A). One option is a module consisting of a single material transfer entering the state variable, with the rate of material transfer represented as a decreasing proportion (calculated in the auxiliary variable) of the level of the state variable. Thus, as the level of the state variable increases, the rate of growth decreases directly.

2.2.6 **Time-lagged oscillations, negative feedback**

There are a variety of modules that produce time-lagged oscillations. A common characteristic of many of these modules is that the negative feedback loop is based on a previous state of the system (Fig. 2.1A). The module presented here is the same as the logistic growth module, except that the negative feedback is based on a previous level of the state variable. The time lag in the negative feedback loop causes oscillations in the state variable, with longer time lags causing oscillations of greater amplitude.

2.3 **One state variable, two material transfers**

2.3.1 **Linear growth or decline, no feedback**

This module produces a behavior similar to the modules of linear growth and linear decline that have a single material transfer. Because both material transfers are constant, the state variable will increase linearly, decrease linearly, or remain in equilibrium forever depending on the relative sizes of the two material transfers (Fig. 2.1B).

2.3.2 **Exponential growth or decline, positive feedback**

This module produces behavior similar to the modules of exponential growth and exponential decline that have a single material transfer. Because both material transfers are constant proportions of the level of the state variable, the state variable will increase exponentially, decrease exponentially, or remain in equilibrium forever depending on the relative sizes of the two proportions controlling the two material transfers (Fig. 2.1B).

2.3.3 Logistic growth, negative feedback

There are a variety of modules that produce sigmoid growth. One option includes modules consisting of one material transfer entering and one material transfer leaving the state variable, in which, as the level of the state variable increases, inflow is represented as a decreasing proportion of the level of the state variable and/or outflow is represented as an increasing proportion of the level of the state variable (Fig. 2.1B). The common characteristic of these modules is the negative feedback loop from the state variable to rate of growth. Thus, as the level of the state variable increases, the rate of growth decreases as a result of proportionally decreasing the rates of inflow and proportionally increasing the rates of outflow.

2.4 Two state variables, two material transfers

2.4.1 Renewable state variable constrained by a nonrenewable state variable, unsustainable growth

This module consists of one (renewable) state variable with an inflow and one (nonrenewable) state variable with an outflow (Fig. 2.1C). The inflow of the first state variable is similar to the inflow of the logistic growth module, except the negative feedback is based on the level of the second state variable rather than on its own level. The outflow of the second state variable is similar to the outflow of the exponential decline module. The growth of the first state variable is unsustainable, reaching a plateau when the second state variable becomes essentially zero. (Note that an exponentially declining value asymptotically approaches, but never reaches, zero.)

2.5 Two state variables, three material transfers

2.5.1 Renewable state variable constrained by a nonrenewable state variable, unsustainable growth and collapse

This module consists of one (renewable) state variable with an inflow and an outflow and one (nonrenewable) state variable with an outflow (Fig. 2.1D). This module is similar to the previous one, except that the first state variable has an outflow, which is similar to outflow of the exponential decline module. Thus, the growth of the first state variable is not only unsustainable, but the level of the state variable will decline after the level of the second state variable passes below a threshold. This module produces the classic "overshoot and collapse" dynamics characteristic of the exploitation of a nonrenewable resource.

2.6 Two state variables, four material transfers

2.6.1 Renewable state variable constrained by a renewable state variable, oscillations

This module consists of two (renewable) state variables, each with an inflow and an outflow (Fig. 2.1E). The first state variable is similar to the first state variable in the previous module. However, the second (now renewable) state variable now has an inflow, which is similar to the inflow of the exponential growth module. Also, the outflow of the second state variable now depends on the level of the first state variable. As the level of the first state variable decreases, the outflow of the second state variable also decreases. This structure produces oscillations in the levels of both state variables, with the growth and decline of the first state variable following the growth and decline of the second state variable. Thus, this module also produces "overshoot and collapse" dynamics in the short term, however, the collapse is followed by recovery.

Worthy of note when comparing these modules with two state variables representing combinations of renewable and nonrenewable state variables, is that nonrenewable resources are stock-limited whereas renewable resources are flow-limited.

2.7 Multiple state variables and material transfers
2.7.1 "Main chain" of state variables, time-lagged oscillations

This module consists of a series of state variables with a material transfer entering the first state variable from a source, a material transfer leaving the last state variable to a sink, and material transfers connecting the first state variable to the second, the second to the third, and the third to the fourth (Fig. 2.1F). The rate of inflow to the first state variable is determined by a driving variable, which, in this example, is calculated as a sine wave. All outflows are equal to the level of the state variable that they are leaving. The level of each state variable oscillates over time in response to the sinusoidal fluctuations of the driving variable. However, a one-unit time lag between fluctuations in the inflow and fluctuations in the outflow is introduced by each successive state variable in the chain. That is, a given change in the driving variable at time t, which determines inflow to the first state variable during time t to $t+1$, is reflected in the level of the first state variable at time $t+1$, in the level of the second state variable at time $t+2$, in the level of the third state variable at time $t+3$, and in the level of the fourth state variable at time $t+4$, and thus in the outflow from the fourth state variable during the time $t+4$ to $t+5$.

2.8 Module applications

There is an almost endless variety of applications of these modules in ecology and natural resource management, as well as in economics, physics, and sociology. In fact, an interesting exercise is to ask a multidisciplinary group of specialists to identify module applications in their particular areas of specialization. The diversity of fundamental processes that exhibit the same general dynamics controlled by the same general processes demonstrates the universality of these modules and the power of the systems perspective in problem-solving. For example, just a few applications in basic population and community ecology might include (1) exponential growth, representing population growth in an unlimited environment; (2) logistic growth, representing population growth in a limited environment; (3) overgrowth and collapse, representing overgrazing and herbivore population crash; (4) oscillations, representing predator-prey cycles; (5) time lags, representing dynamics of age-structured populations. Rather than providing a long list of specific examples here, we encourage readers to supplement our brief list with examples from their own areas of specialization.

The modeling
process

Overview of procedures

3

Chapter outline

A variety of schemes for applying the systems approach to ecological modeling has been suggested over the years (e.g., Kitching, 1983; Grant, 1986; Ford, 1999; Grant and Swannack, 2008). These schemes differ with regard to details (number of steps, names of steps), but all are based on the same underlying general systems theory. Our preference, which has been shaped by application of the systems approach for modeling a variety of problems in ecology and natural resource management over the past four decades, is to identify four fundamental phases in the process of developing and using a systems model: (1) conceptual model formulation, (2) quantitative model specification, (3) model evaluation, and (4) model application (Fig. 3.1). The remainder of this chapter provides a brief overview of each of these four basic phases, each of which will be examined in more detail in Part 2 of the book, as well as a few examples of their application to model simple hypothetical systems.

3.1 Conceptual model formulation

The goal of the first phase of the modeling process is to develop a conceptual, or qualitative, model of the system-of-interest (Fig. 3.1A). Based on the objectives of the modeling project, we decide which components in the real-world system

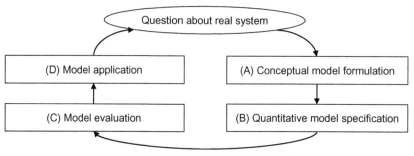

FIG. 3.1

Four phases in the modeling process: (A) conceptual model formulation, (B) quantitative model specification, (C) model evaluation, and (D) model application.

should be included in our system-of-interest, and how they should be related to one another. We represent these components and their relationships, which collectively form our conceptual model, diagrammatically using symbols that indicate the specific nature of the relationships. We also sketch patterns of behavior that we expect our model will exhibit, often in terms of the general temporal dynamics of key system components. This is the most inductive, creative, synthetic, constructive, "artistic" phase of the modeling process during which we try to decide what to include in this abstraction, this "caricature," of reality which is the model, and how to include it.

3.2 Quantitative model specification

The goal of the second phase of systems analysis is to develop a quantitative model of the system-of-interest (Fig. 3.1B). This basically involves translating our conceptual model, which is represented diagrammatically and using words, into a series of mathematical equations that collectively form the quantitative model. This translation, or quantification, is based on the consideration of various types of information about the real system. We then solve all of the model equations, or instruct a computer to solve them, each time step over the entire period of simulated time in which we are interested. That is, we run the baseline simulation. This is a more deductive, mechanical phase of the modeling process during which we use mathematical analysis to a construct a "deductive engine," which is a model that simulates numerically the ideas embodied in the conceptual model.

3.3 Model evaluation

The goal of the third phase of systems analysis is to evaluate the usefulness of the model in meeting our objectives (Fig. 3.1C). This process, which is the subject of much debate, commonly is referred to as "model validation" and often erroneously focuses on comparison of model predictions with real-system observations as the only validation criterion. We prefer the concept of "model evaluation" based on consideration of a broad array of different aspects of model structure and behavior that make it potentially useful. We may place more emphasis on interpretability of relationships among components within the model or on predictive capabilities of the model, depending on our objectives. Often, we are interested in determining how sensitive model predictions are to the uncertainties with which we have represented certain aspects of the model. This also is a more deductive, mechanical phase of the modeling process, although selection of the aspects of the model to be examined during sensitivity analysis involves an artistic component, particularly for very large models.

3.4 Model application

The goal of the final phase of systems analysis is to answer the questions that were identified at the beginning of the modeling project (Fig. 3.1D). This involves designing and simulating the same experiments with the model that we would conduct in the real system to answer our questions. We also analyze, interpret, and communicate simulation results using the same general procedures that we would use for real-world results. This also is in large part a deductive, mechanical phase of the modeling process in that we use our model, our "deductive engine," to generate answers to our questions about the system-of-interest. Arguably, however, the effective communication of simulation results to the appropriate audiences involves an artistic component.

3.5 Iteration of phases

The four phases of systems analysis are highly interconnected. Although theoretically we may think of the process as proceeding sequentially in the indicated order (Fig. 3.1), in practice, we may cycle through several phases more than once. During any phase, we may find that we have overlooked or misrepresented an important system component or process and need to return to an earlier phase, often to conceptual model formulation or quantitative model specification.

During model evaluation in particular, we examine the model to detect any inadequacies that may require us to cycle back to earlier phases. Discovery of such inadequacies in the model during its development usually provides additional insight into the dynamics of the system-of-interest and is an important benefit of modeling. We might characterize this iteration of phases as a process during which we alternate between inductive and deductive thinking and during which we often are surprised by the gaps in our understanding that such attempts at synthesis usually reveal. We will demonstrate this iterative nature of model development in more detail when we take a candid look at the practical application of simulation modeling in ecology in Chapter 9.

3.6 A few simple examples

To illustrate the application of these four fundamental phases in the process of developing and using a systems model, we offer a few (apparently very simple) hypothetical examples. The first deals with a bathtub and the rate at which it fills up with water. The second deals with a nonrenewable natural resource and the rate at which it is extracted from the ground. The third deals with an animal population and the rate at which it can recover from a large mortality event. In each case, of course, we begin with a question about the real-world system, about our system-of-interest (Fig. 3.1).

3.6.1 How long will it take the bathtub to fill up?

We have observed that our bathtub fills up at different rates depending on how far we open the faucet. We wonder how many minutes it will take for the bathtub to fill up with water if we open the faucet one turn versus two turns versus three turns. We sketch a diagram of our system-of-interest using the symbols introduced in Chapter 2 (Section 2.1) (Fig. 3.2). That is, we formulate our conceptual model (Fig. 3.1A). From our conceptual model, we see that, in order to answer our question, in order to quantify our conceptual model, we need to know this size of the bathtub and the rates of water inflow resulting from one, two, and three turns of the faucet.

We gather data that suggest the tub holds 150 glasses of water, and the rates of water inflow resulting from one, two, and three turns of the faucet are 5, 10, and 15 glasses per minute, respectively. Thus, the rate of water inflow equals five times the number of faucet turns. Following the style of the simple numerical example presented in Chapter 1 (Section 1.2.5), we write equations that specify quantitatively our model (Fig. 3.1B).

$$Tub_{t+1} = Tub_t + (inflow)\,\Delta t \tag{3.1}$$

$$inflow = f(faucet\ turns) \tag{3.2}$$

In Eq. (3.1), Δt represents 1 minute. To quantify the functional relationship represented in Eq. (3.2), we write

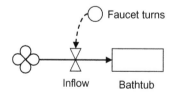

FIG. 3.2

Conceptual model representing how our bathtub fills up at different rates depending on how far we open the faucet (one turn, two turns, three turns).

$$inflow = 5 \times faucet\ turns \qquad (3.3)$$

We note that $Tub_0 = 0$, and that *faucet turns* can equal 1, or 2, or 3. We then solve the model equations each time step over a period of simulated time long enough for the bathtub to fill up, assuming the faucet has been opened 1 turn. That is, we run the baseline simulation. Thus, for the first minute

$$inflow = 5 \times 1 = 5 \qquad (3.4)$$

$$Tub_1 = 0 + (5) = 5 \qquad (3.5)$$

and for the second minute

$$inflow = 5 \times 1 = 5 \qquad (3.6)$$

$$Tub_2 = 5 + (5) = 10 \qquad (3.7)$$

Repeating the same sequence of calculations, we find that it takes 30 minutes for the bathtub to fill up under the baseline conditions.

To evaluate the usefulness of the model in meeting our objectives (Fig. 3.1C), we consider model structure as well as model behavior under the baseline conditions. Model structure seems reasonable in that, given the objective of the model, we only are interested in increases in the amount of water in the bathtub. Of course, since the model has no outflow, we are assuming that there are no leaks in the bathtub. Model behavior also seems reasonable in that the amount of water in the bathtub should increase linearly, since the rate of inflow is constant. Or, if not constant, we are assuming that any variations in flow rate are inconsequentially small. (A word of caution regarding model evaluation is in order here. Although it would be a distraction at this point to belabor the evaluation of such a simple model that addresses such a straightforward question, the importance of the process of model evaluation increases exponentially with the complexity of the model and with the significance to users of the simulation results. We discuss the topic of model evaluation at length in Chapters 6 and 10.)

To apply the model to answer the questions that we identified at the beginning of the modeling project (Fig. 3.1D), we design and execute a simulated experiment in which we compare the lengths of time it takes the bathtub to fill up with the faucet open 1 versus 2 versus 3 turns. Repeating the same sequence of calculations illustrated in Eqs. (3.1)–(3.7), we find that it takes 30, 15, and 10 minutes for the bathtub to fill up assuming 1, 2, and 3 faucet turns, respectively (Fig. 3.3). (Another word of caution is in order here. Discourses on the principles of experimental design and methodologies of data analysis are beyond the scope of this book. We touch briefly on statistical analysis within the context of simulation modeling in Chapter 5 (Quantitative model specification), Chapter 6 (Model evaluation), and Chapter 7 (Model application). For our purposes, suffice it to say that simulation results, that is, simulated data, are indeed data, and hence subject to the same assumptions and requirements of field or laboratory data when considering statistical methods of analysis.)

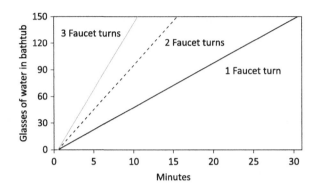

FIG. 3.3

Simulated lengths of time it takes the bathtub to fill up with the faucet open 1 versus 2 versus 3 turns.

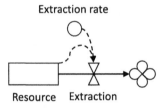

FIG. 3.4

Conceptual model representing how much longer a nonrenewable natural resource will last under different rates of annual extraction.

3.6.2 How long will the nonrenewable natural resource last?

We have observed that a nonrenewable natural resource that we extract from the ground is becoming alarmingly less abundant. The most recent study estimated that there are about 100,000 units of the resource remaining. Current regulations limit the amount extracted annually to 10% of the estimated amount remaining. We wonder how many more years the resource would last if we decrease the allowed annual extraction rate from 10% to 5% or 2.5% of the estimated amount remaining. We formulate our conceptual model based on our observations (Fig. 3.4), and we specify quantitatively our model based on the available data.

$$Resource_{t+1} = Resource_t - (extraction)\, \Delta t \qquad (3.8)$$

$$extraction = f(Resource_t, extraction\ rate) \qquad (3.9)$$

In Eq. (3.8), Δt represents 1 year. To quantify the functional relationship represented in Eq. (3.9), we write

$$extraction = extraction\ rate \times Resource_t \qquad (3.10)$$

We note that $Resource_0 = 100,000$, and that *extraction rate* can equal 0.10, or 0.05, or 0.025. We then solve the model equations each time step over a period of simulated time long enough for the nonrenewable natural resource to become diminishingly small, which we define arbitrarily as being less than 1 unit. We assume a baseline annual extraction rate of 10%, which we will represent as a proportion (0.01). Thus, for the first year

$$extraction = 0.10 \times 100,000 = 10,000 \qquad (3.11)$$

$$Resource_1 = 100,000 - (10,000) = 90,000 \qquad (3.12)$$

and for the second year

$$extraction = 0.10 \times 90,000 = 9000 \qquad (3.13)$$

$$Resource_2 = 90,000 - (9000) = 81,000 \qquad (3.14)$$

Repeating the same sequence of calculations, we find that it takes 110 years for the nonrenewable natural resource to decrease to less than 1 unit.

To evaluate the usefulness of the model in meeting our objectives, as before, we consider model structure as well as model behavior under the baseline conditions. Model structure seems reasonable in that, given the objective of the model, we only are interested in decreases in the remaining amount of nonrenewable natural resource. Model behavior also seems reasonable in that the amount of nonrenewable natural resource remaining should decrease exponentially (the rate of outflow is constant proportion of a continually decreasing amount remaining). As with the bathtub example, of course, there are a series of implicit assumptions pertaining to model structure and functional relationships that we might list. For example, that the natural resource is nonrenewable for all practical purposes within the time frame of our interests.

To apply the model to answer our original questions, we design and execute a simulated experiment in which we compare the lengths of time it takes for the nonrenewable natural resource to decrease to less than 1 unit, assuming annual rates of extraction of 10%, 5%, and 2.5%, which we will represent as proportions (0.1, 0.05, 0.025). Repeating the same sequence of calculations illustrated in Eqs. (3.8)–(3.14), we find that it takes 110, 225, 455 years, respectively (Fig. 3.5).

3.6.3 How fast will the population recover?

We have observed that an animal population which inhabits an oceanic island has become drastically reduced in size following a large hurricane. We wonder how many years it will take for the population to recover to its prehurricane size. The most recent study estimated that there only are about 10 animals per km^2 remaining in the population. Available data indicate that population size on the island before the hurricane was about 400 animals per km^2. Data also suggest that the (nonhurricane-related) annual death rate historically has been 50%, and that annual per capita birth

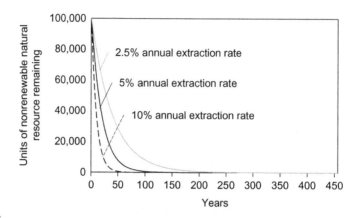

FIG. 3.5

Simulated lengths of time it takes for the nonrenewable natural resource to decrease to less than 1 unit, assuming annual rates of extraction of 10%, 5%, and 2.5%.

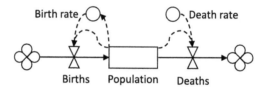

FIG. 3.6

Conceptual model representing how fast an oceanic island animal population can recover to its prehurricane size following a large hurricane.

rate decreases linearly from a maximum of 1 to a minimum of 0 as population size increases from essentially zero to 800 per km^2. As before, we formulate our conceptual model based on our observations (Fig. 3.6), and we specify quantitatively our model based on the available data.

$$Pop_{t+1} = Pop_t - (births - deaths)\,\Delta t \qquad (3.15)$$

$$births = f(Pop_t, birth\ rate) \qquad (3.16)$$

$$deaths = f(Pop_t, death\ rate) \qquad (3.17)$$

In Eq. (3.15), Δt represents 1 year. To quantify the functional relationships represented in Eqs. (3.16) and (3.17), we write

$$births = birth\ rate \times Pop_t \qquad (3.18)$$

$$deaths = death\ rate \times Pop_t \qquad (3.19)$$

$$birth\ rate = 1 - (0.00125 \times Pop_t) \qquad (3.20)$$

$$death\ rate = 0.50 \qquad (3.21)$$

We note that $Pop_0 = 10$. We then solve the model equations each time step over a period of simulated time long enough for the population to recover to its prehurricane size. Thus, for the first year

$$births = (1 - 0.00125 \times 10) \times 10 = 9.875 \qquad (3.22)$$

$$deaths = 0.50 \times 10 = 5 \qquad (3.23)$$

$$Pop_1 = 10 + (9.875 - 5) = 14.875 \qquad (3.24)$$

and for the second year

$$births = (1 - 0.00125 \times 14.875) \times 14.875 = 14.598 \qquad (3.25)$$

$$deaths = 0.50 \times 14.875 = 7.44 \qquad (3.26)$$

$$Pop_2 = 14.875 + (14.598 - 7.44) = 22.036 \qquad (3.27)$$

Repeating the same sequence of calculations, we find that it takes about 19 years for the population to recover to approximately (399.508 animals per km^2) its prehurricane size.

To evaluate the usefulness of the model in meeting our objectives, once again, we consider model structure as well as model behavior under the baseline conditions. Model structure seems reasonable in that, given the objective of the model and the data available, we are interested in increases in population size resulting from the net difference between annual births and deaths. Model behavior also seems reasonable. Based on population growth theory, we know that populations growing in a limited environment should exhibit sigmoidal (S-shaped) growth. That is, population size should increase at an increasing rate until the population reaches one-half of the carrying capacity of the environment, and then should increase at a decreasing rate until it reaches carrying capacity. As with the previous two examples, there are a series of implicit assumptions pertaining to model structure and functional relationships that we might list. For example, that we are dealing with a closed population, that is, there is no immigration nor emigration.

To apply the model to answer our original questions, note that, in this case, results of the baseline simulation suffice. That is, the baseline simulation results suggest that the population can recover to its prehurricane size in about 19 years (Fig. 3.7).

3.6.4 Some final comments

In these simple examples, we have not sketched patterns of behavior that we expect our model will exhibit, which we mentioned as a part of the conceptual model formulation phase (Section 3.1), nor have we illustrated a sensitivity analysis, which we mentioned as a part of model evaluation phase (Section 3.3), nor have we demonstrated the iteration of the various phases (Section 3.5). Examples of these three items related to our population model might be as follows. We probably would have sketched an expected pattern of post-hurricane population growth similar to our simulation results in Fig. 3.7. Given the negative feedback of population size on

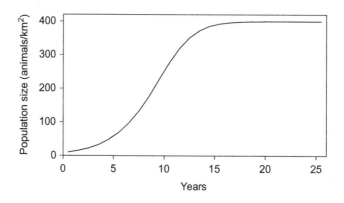

FIG. 3.7

Simulated length of time it will take an oceanic island animal population to recover to its prehurricane size following a large hurricane.

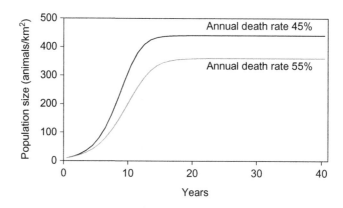

FIG. 3.8

Simulated length of time it will take an oceanic island animal population to recover to its prehurricane size following a large hurricane, assuming annual death rates of 45% and 55%, respectively.

population birth rate, we would have expected logistic (S-shaped) population growth toward carrying capacity (400 animals per km^2). We might conduct a sensitivity analysis on our population model to address the following question. How much would our estimate of the time it takes for the population to recover to its prehurri-cane size be changed if the annual death rate was 45%, or 55%, rather than 50%? Or, in other words, how sensitive is our estimate of recovery time to uncertainty in our estimate of the parameter representing annual death rate? Repeating the same sequence of calculations illustrated in Eqs. (3.22)–(3.24), assuming annual death rates of 45% and 55%, we find that it takes the population about 18 and 21 years, respectively, to reach carrying capacity (Fig. 3.8). Within the context of our original

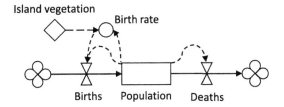

FIG. 3.9

Revised (to include explicit representation of the island habitat) conceptual model representing how fast an oceanic island animal population can recover to its prehurricane size following a large hurricane.

concern about the population, this may or may not seem to be an important difference. However, we note that the carrying capacities reached when assuming annual death rates of 45% and 55% are about 440 and 360 animals per km^2, respectively, rather than 400 animals per km^2. This may raise doubts about the prehurricane estimate of 400 animals per km^2, or perhaps suggest the need to represent the habitat characteristics on the island explicitly in the model.

To demonstrate the iteration of the various modeling phases, we might pursue the idea of representing the island habitat explicitly in the model. This would lead us to change our conceptual model to include the island habitat, perhaps as a variable whose value would increase over time to represent the posthurricane recovery of island vegetation, thus increasing annual birth rate of the population (Fig. 3.9). We then would need to parameterize the new relationship between island vegetation and annual birth rate, re-evaluate the model, probably including a new sensitivity analysis, and reapply the model to address our initial question.

Conceptual model formulation

Chapter outline

4.1 Overview of the process

The goal of the first phase of the simulation modeling process is to develop a conceptual, or qualitative, model of the system-of-interest (Fig. 4.1). Based on a clear statement of the objectives of the modeling project, we abstract from the real system those components that we think should be considered to address our questions. By including these components within our model and excluding all others, we bound the system-of-interest. Next, we categorize model components depending on their specific roles in describing system structure and identify specific relationships among components that generate system dynamics. We then formally represent the resulting conceptual model, usually as a box-and-arrow diagram indicating points of accumulation of material (boxes), such as individuals, energy, biomass, nutrients, or some other entity of interest, and routes by which the material flows within the system (arrows).

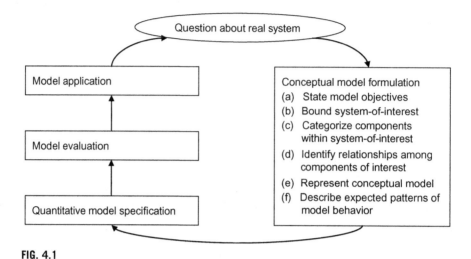

FIG. 4.1

Steps within Phase I of the simulation modeling process: conceptual model formulation.

Finally, we describe expected patterns of model behavior, most often as graphs representing changes in values of important variables within the system over time.

   In many respects, conceptual model formulation is the most intellectually challenging and "artistic" phase of the modeling process. The best basis for the many difficult, and often highly subjective, decisions that must be made regarding choice of model components is a thorough familiarity with the real system. Prior modeling experience also is an asset. There are two general approaches to identifying model components. One approach makes the initial choice of components as simple as possible and subsequently adds critical components that were overlooked. The other approach includes initially all components that possibly could have any importance and then delete superfluous ones. Theoretically, the end product of either approach should be a conceptual model that is no more complex than is absolutely necessary to address our interests. As we will see in Chapter 9, in practice, it is better to begin with the simplest model possible.

4.2 **State model objectives**

We begin with a clear statement of the purpose of our model in terms of a problem to be solved or a question to be answered. Questions may arise from general observations of a system, as is the usual case in scientific inquiry, or may be imposed by the practical necessity of evaluating proposed management schemes. We also must specify the criteria the model must meet to be considered useful, given its purpose. Criteria may require that the model have theoretically reasonable structure and interpretable functional relationships, that model behavior corresponds well with expected patterns of real-system behavior, and/or that model projections correspond well with data from the real system. The relative importance of each type of criteria depends on the purpose of the model. Finally, we must describe the context within which we intend the model operate. Context includes all the restrictive assumptions we must make for the model to be a useful representation of the real system. Obviously, an exhaustive list of these assumptions is impossible, and how to distinguish between those assumptions that can be left implicit (the earth will continue to revolve around the sun) and those that should be stated explicitly (no changes in general climatic conditions) will remain enigmatic. Nonetheless, we should give serious thought to the manner in which we can communicate most effectively the essence of this context to those who will use the model or its results.

Because model objectives provide the framework for model development, the standard for model evaluation, and the context within which simulation results will be interpreted, this is arguably the most crucial step in the entire modeling process. Yet, surprisingly, this step usually receives far less attention than its importance warrants.

Often, our initial formulation of an objective is too broad to address directly and thus is of little use in guiding model development. As a general rule, objectives that begin with "to understand ..." need to be stated more specifically. For example, recalling the simple examples with which we concluded Chapter 3, compare the following pairs of objective statements. (a) Understand how the bathtub fills up when we turn on the faucet. (b) Determine how many minutes it will take for the bathtub to fill up with water if we open the faucet one turn versus two turns versus three turns. (a) Understand how much longer the nonrenewable resource will last. (b) Determine how many more years the nonrenewable resource would last if we decrease the allowed annual extraction rate from 10% to 5% or 2.5% of the estimated amount remaining. (a) Understand how the population will recover after the hurricane. (b) Determine how many years it will take for the population to recover to its prehurricane size. In each case, the first objective statement provides a general idea of the problem, but does not specify exactly how we will approach the problem. It remains unclear exactly what will we will we monitor and exactly what we will manipulate, what model parameters we will alter. The second objective statement indicates exactly what will we monitor and, in the first two examples, what will we manipulate to address our questions. (In the third example, results of the baseline simulation sufficed to answer our original question. That is, we did not need to alter model parameters to address our question.)

4.3 **Bound system-of-interest**

Bounding the system-of-interest consists of separating those components that we think should be included within the system-of-interest from those that we think should be excluded. We do not want an unnecessarily complex model, but likewise we do not want to exclude components that might be critical to the solution of our problem. Obviously, this step in conceptual-model formulation is highly subjective, and it is common to define and redefine system boundaries several times as we are developing our conceptual model. Consider for a moment some other ways that we might have bounded the systems associated with the simple examples with which we concluded Chapter 3.

In the bathtub example (Fig. 3.2), we assumed that there were no leaks in the bathtub. Suppose that, after further consideration, we decided we should include a leak which was due to a faulty plug that did not seal the drain completely. In the non-renewable natural resource example (Fig. 3.4), we assumed that the natural resource was, in fact, nonrenewable for all practical purposes within the time frame of our interests. Suppose that we decided we should allow for the possibility of discovering new sources of the resource due to improved technologies. In the population example (Fig. 3.6), we assumed that we were dealing with a closed population, that is, there was no immigration nor emigration. Suppose that, although emigration still seemed unlikely, we decided that we should include immigration. So how might we modify our conceptual models to represent these redefined system boundaries, and these new aspects of the systems-of-interest? How might we represent a leak in the bathtub, the discovery of new sources of the resource, and immigration to the population? And, a related question, which may have occurred to you, why do we represent some system components in our conceptual models as squares and some as circles, and why do we represent some connections among components with solid lines and some with dashed lines? These questions lead us to the next step in conceptual model formulation.

4.4 **Categorize components within system-of-interest**

Once the system-of-interest has been bounded initially by separating those components that we think should be included within the system from those that we think should be excluded, and by identifying particular attributes of system components that are of interest, we proceed to step 3 of conceptual-model formulation, categorizing components within the system-of-interest. System components do not all serve the same purpose in a model. Certainly, they all represent important aspects of the system-of-interest, but there may be as many as seven fundamentally different categories of system components: (1) state variables, (2) material transfers, (3) sources and sinks, (4) information transfers, (5) constants, (6) driving variables, and (7) auxiliary variables (Fig. 4.2). Most of the symbols in Fig. 4.2 can be traced back to Forrester (1961) and subsequently were added to (Innis, 1979) and have been adapted (e.g., Grant, 1986; Grant et al., 1997; Grant and Swannack, 2008).

4.4.1 **State variables**

State variables (represented by boxes in Fig. 4.2) represent points of accumulation of material within the system. The accumulation of some material in which we are particularly interested. The bathtub (Fig. 3.2) is an accumulation of glasses of water,

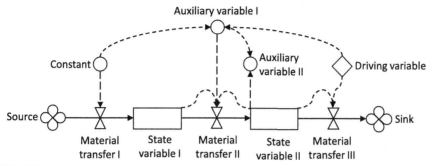

FIG. 4.2

Symbols used to construct conceptual model diagrams indicating all permissible connections among model components. The *diamond shape* represents a driving variable. The *boxes* represent state variables. The *solid lines* represent material transfers. The *dashed lines* represent information transfers. The *circles* represent constants if they receive no information transfers, or auxiliary variables if they do receive information transfers. The clouds represent sources and sinks.

the nonrenewable resource (Fig. 3.4) is an accumulation of, say, fossil fuel in metric tons, the population (Fig. 3.6) is an accumulation of individual animals per km^2. Other examples of state variables might include the kcal of energy accumulated in plants, herbivores, and carnivores, if we are interested in energy flow through an ecosystem, or the kilograms of nutrients accumulated in plants, herbivores, carnivores, and an abiotic component, if we are interested in nutrient cycling in an ecosystem.

4.4.2 Material transfers

The accumulation of material in a state variable may increase or decrease via material transfers (represented by solid lines in Fig. 4.2). A material transfer represents the movement of material (1) from outside the system into a state variable, (2) between two state variables, or (3) from a state variable out of the system. The flow of glasses of water into the bathtub (Fig. 3.2) is a material transfer. The movement of metric tons of fossil fuel out of the nonrenewable resource (Fig. 3.4) is a material transfer. Both the births and deaths in the population model (Fig. 3.6) are material transfers. They represent the "movement" of animals into and out of the population. Other examples of material transfers might include the movement, or flow, of kilocalories of energy into and out of the plants, herbivores, and carnivores in an ecosystem, or the movement of kilograms of nutrients among plants, herbivores, carnivores, and an abiotic component in an ecosystem. You might stop for a moment to sketch what these two conceptual models might look like.

Note that material transfers obey the laws of thermodynamics. If one unit of material leaves a donor state variable, one unit of material arrives at the receiving state variable. Also note that units of measure associated with material transfers always include the units of measure of the state variable(s) to which they are connected, plus a time unit. For example, glasses of water per minute, metric tons of fossil fuel per year, and number of animals per year. Thus two state variables connected by a material transfer must have the same units of measure. Specifying the appropriate time units (second, year, century) occurs in the second phase of systems analysis, but, looking ahead a bit, the basic assumption we make is that the rate of material transfer is constant over the time unit we specify. If we feel uncomfortable with this assumption, that is, if we think the rate of movement may change during the specified time period in an important way, we need to shorten the time unit. (Discussion of instantaneous and finite rates of change, and differential and difference equations, would be a distraction at this point.)

Returning to the question of how we might modify our conceptual models to represent a leak in the bathtub, the discovery of new sources of the resource, and immigration to the population, you probably have noticed that all of these additional aspects of the systems-of-interest represent the movement of material into or out of the system, into or out of a state variable. Thus, we might modify the diagrammatic representations of our conceptual models appropriately by adding material transfers originating from sources and terminating in sinks (Fig. 4.3).

4.4.3 **Sources and sinks**

Sources and sinks (represented by "clouds" in Fig. 4.2) represent origination and termination points, respectively, of material transfers into and out of the system. By definition we are not interested in the level of accumulation of material within sources and sinks, we are interested only in the rates at which material enters and leaves the system. Note that closed systems, or systems-of-interest that we conceptualize as being closed, have no sources or sinks. A model of nutrient cycling between abiotic and biotic components within an ecosystem might be an example. Also worthy of note is that "closed" within a systems modeling context means that the total amount of material in the system-of-interest does not change over time. However, the distribution of that material among various state variables may, and probably does, change over time. (And a cautionary note for population biologists. A "closed" population, implying no immigration nor emigration, is an "open" system from a systems modeling point of view as long as there are births or deaths. That is, as long as material in the form of individuals, or biomass, enters or leaves the system.)

4.4.4 **Information transfers**

Information transfers (represented by dashed lines in Fig. 4.2) represent the use of information about the state of the system to control the change of state of the system. This concept of the rate of change in the state of the system being a function of the current state of the system is a key concept in systems simulation. It follows that

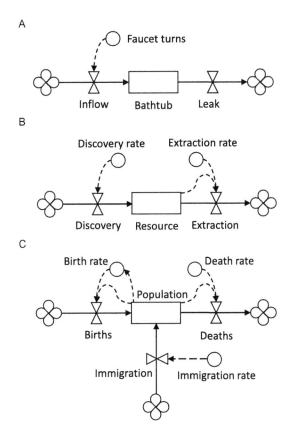

FIG. 4.3

Modifications to the conceptual models of (A) the bathtub, now representing a leak; (B) the resource, now representing the discovery of new sources; and (C) the population, now representing immigration.

conceptual models of all but the simplest of systems (such as the original conceptual model for our bathtub in Fig. 3.2) will contain at least one information transfer originating from a state variable whose path can be traced to a material transfer. That is, information about the current state of the system, which is embodied in the values of the state variables, will be used to determine the change in the state of the system, which results from adding and subtracting the values of the material transfers to and from the values of the state variables to which they are connected. The path of an information transfer from a state variable to a material transfer may be direct, as in the simple examples we have seen thus far, or indirect, as illustrated by the conceptual model in Fig. 4.4. The conceptual model in Fig. 4.4 represents the biomass dynamics of a hypothetical plant-fruit-frugivore (fruit-eater) system. Plant growth is

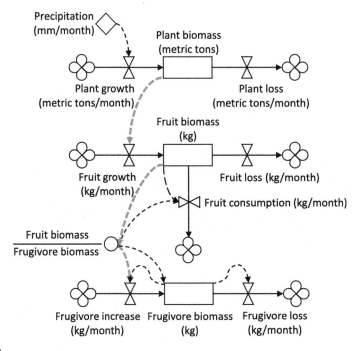

FIG. 4.4

Conceptual model representing the biomass dynamics of a hypothetical plant-fruit-frugivore (fruit-eater) system, which illustrates an indirect path (highlighted *dashed lines*) connecting an information transfer from a state variable (plant biomass) to a material transfer (increase in frugivore biomass). See text for details.

a function of precipitation. Fruit production is a function of plant biomass. Loss of fruit biomass via consumption by frugivores is a function of the ratio of fruit biomass to frugivore biomass, which represents relative food availability. Increase in frugivore biomass also is a function of the ratio of fruit biomass to frugivore biomass. But note that increase in frugivore biomass is connected indirectly to plant biomass via a series of information transfers, which are highlighted in Fig. 4.4. Unlike material transfers, the units of measure we use to represent information can change as the information passes from one system component to another, as can the numerical value assigned to the information being transferred, as long as the appropriate conversion factors are used. In Fig. 4.4, information on plant biomass (metric tons) is used to calculate fruit production (kg/month). Information on fruit biomass (kg) and frugivore biomass (g) is converted into a ratio (kg/g) and used to calculate frugivore increase (g/month). We refer collectively to the information transfers within a system as the information network, which connects not only state variables and material transfers but also driving variables, constants, and auxiliary variables.

4.4.5 **Constants**

Constants (represented by circles with no dashed lines entering in Fig. 4.2) are numerical values describing important characteristics of a system that do not change, or that can be represented as unchanging, under all of the conditions encountered in any given scenario simulated by the model. In our modified conceptual models for the bathtub, the resource, and the population (Fig. 4.3), faucet turns (Fig. 4.3A), extraction rate and discovery rate (Fig. 4.3B), and death rate and immigration rate (Fig. 4.3C) all have been defined as constants (birth rate is an auxiliary variable, described in Section 4.4.7). Note that our definition of constants is tightly linked to the manner in which we have conceptualized our system-of-interest. Model components that can be defined as constants at one level of detail and temporal scale may not be defined as constants at finer levels of detail or longer temporal scales. Note also that we need not, and usually do not, represent all of the mathematical constants in our model as constants in our conceptual model. Quite likely there will be coefficients that appear in equations throughout the model that are not, by themselves, conceptually important.

4.4.6 **Driving variables**

Driving variables (represented by a diamond shape in Fig. 4.2) affect, but are not affected by, the rest of the system. As with constants, our definition of driving variables is tightly linked to the manner in which we have bounded our system-of-interest. In the conceptual model in Fig. 4.4, precipitation is represented as a driving variable, it affects plant growth but is not affect by plants or anything else in the system. Factors such as environmental temperature and precipitation, which often are represented as driving variables, also may be represented as constants, if they do not change during the course of any given simulation.

4.4.7 **Auxiliary variables**

Auxiliary variables (represented by circles with dashed lines entering in Fig. 4.2) most commonly represent processes or concepts in the system-of-interest that we wish to indicate explicitly, which otherwise would be implicit in the information transfers among model components (constants, driving variables, state variables, material transfers). Such auxiliary variables might be viewed as intermediate steps in determining a rate of material transfer or the value of another auxiliary variable. For example, birth rate in our population model (Fig. 4.3C) and the ratio of fruit biomass to frugivore biomass in our plant-fruit-frugivore model (Fig. 4.4) are auxiliary variables, which we might have omitted from our conceptual models. In our population model, we could have connected population directly to births (Fig. 4.3C) and imbedded the quantitative relationship between the population size and per capita birth rate within the births material transfer equation. Likewise, we could have connected fruit biomass directly to frugivore increase (Fig. 4.4) and imbedded the quantitative relationship between the fruit biomass/frugivore biomass ratio within the frugivore increase material transfer. Our choice to include these two auxiliary variables was based on the goal of making the conceptual model as useful to the intended audience as possible. In the case of our population model, we might have wanted to call attention to the density-dependency of per capita birth rates. In the case of our plant-fruit-frugivore model, we might have wanted to emphasize the importance of the relative, as opposed to absolute, availability of food in controlling furgivore increase.

Auxiliary variables also may represent an alternative form of some model component that, although it does not affect system dynamics, is of particular interest to us as a model output. Such auxiliary variables often are used simply to express the value of a model component in different units of measure. For example, we might have used an auxiliary variable to represent the monetary value of fruit biomass. In this case, we would simply transfer information from the fruit biomass state variable to the new auxiliary variable and use the appropriate conversion factor to convert kilograms into monetary units.

4.5 Identify relationships among components of interest

The fourth step of conceptual-model formulation consists of identifying relationships among system components that are of interest. As we have seen in the previous section, there are two ways that system components can be connected: by material transfers or by information transfers (Fig. 4.2). To summarize, material transfers can enter a state variable from a source, connect two state variables, or leave a state variable to a sink. Material transfers obey the laws of thermodynamics. If one unit of material leaves a donor state variable, one unit of material must arrive at the receiving state variable. It follows that units of measure of state variables connected by material transfers must be the same. Units of measure of material transfers must be the same as the state variables to which they are connected, with the addition of a time unit. Information transferred within the system refers to information about current values of state variables, driving variables, constants, and auxiliary variables. This information is transferred for use in determining the rates at which material transfers occur or for calculating specific results, or output auxiliary variables, required of the model. Information transfers can leave state variables, driving variables, constants, and auxiliary variables and can enter material transfers and auxiliary variables. Units of measure of variables affecting a given material transfer or auxiliary variable need not be the same, but, of course, the different units of measure must be manipulated appropriately and must be consistent with the definition of the system components involved.

4.5.1 Submodels

Often, it is useful conceptually to describe larger models in terms of submodels. The manner in which submodels are defined depends on the particular aspects of the model that we want to emphasize. In general, submodels are defined such that they distinguish important groups of system components and processes. Sometimes submodels are defined to emphasize the different types of material that are "flowing" in the system, with each submodel representing the dynamics of a different material and submodels connected to each other solely via information transfers. For example, in our plant-fruit-frugivore model (Fig. 4.4), we might have defined a plant submodel, a fruit submodel, and a frugivore submodel. The material represented in each of these three submodels is different, and the three state variables are not connected by material transfers, only by information transfers. Another example might be submodels defined such that they distinguish physical environmental, ecological, and social (human) system components, or abiotic and biotic system components. These decisions are completely subjective and are based on whether or not we think division into submodels facilitates understanding of the model.

4.6 Represent conceptual model

Formal representation of the conceptual model most commonly takes the form of a box-and-arrow diagram such as those we have been using thus far. As we have seen, such diagrams play an important role in modeling by helping us visualize the "big picture" and by facilitating communication among different people who are interested in a particular system. Although we present this as the fifth step in conceptual model formulation, and indeed the conceptual model diagram might be thought of as the end product of the first phase of systems analysis, diagrammatic representation of the conceptual model usually is concurrent with the earlier steps and aids them greatly. Conceptual model diagrams also provide a framework that facilitates subsequent quantification of the model because equations can be related directly to specific parts of the conceptual model.

A variety of schemes exist for formal representation of the conceptual model. The origin of the particular symbols we use throughout this book, and the concepts they represent, can be traced back (at least) to the "Forrester Diagrams" used for modeling dynamics of industrial systems (Forrester, 1961). We have modified the shapes of the symbols a bit over the years (Grant, 1986; Grant et al., 1997), but only for the sake of convenience in generating them.

4.7 Describe expected patterns of model behavior

We almost always have some expectations concerning patterns of model behavior before ever running the first simulation. We should formally describe these expectations so we can draw upon them (1) as points of reference during model evaluation and (2) to ensure that the model provides the types of projections that allow us to address our questions directly during model application. These expectations are based on all of our current knowledge about the system-of-interest, including, but not limited to, what can be supported by available data. Most often we know more

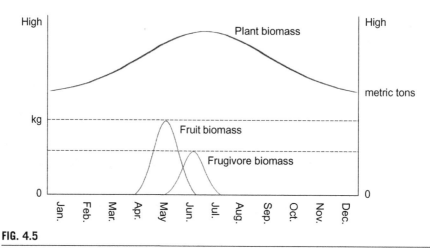

FIG. 4.5

Expected patterns of biomass dynamics of the plant, fruit, and frugivore (fruit-eater) components of the hypothetical plant-fruit-frugivore system.

about relationships among variables within the system-of-interest than can be documented in a rigorous way by data, and almost always there are important aspects of system dynamics for which we simply have no data. We should describe the expected behavior of those variables that most effectively represent this broader knowledge, thus allowing a more extensive evaluation of model behavior than would be possible based solely on data. We also should describe the expected behavior of those variables that most directly represent hypotheses that we want to test. We usually formalize these expectations as graphs representing changes in values of important variables over time. For example, supposing that the objective of our plant-fruit-frugivore model was to estimate the seasonal time lag between fruit production and frugivore population increase, we might sketch the expected patterns of biomass dynamics of the plant, fruit, and frugivore components system. Our expectation might be that plant biomass is relatively stable, that fruit biomass increases rapidly from April to May and then decreases rapidly from May to June, and that frugivore biomass increases and decreases follow fruit biomass dynamics with a 2-month time lag (Fig. 4.5). But we also should note any aspect of system behavior for which we have an expectation. For example, we might expect that the seasonal peak in frugivore biomass would reach only one-half the magnitude of the seasonal peak in fruit biomass (Fig. 4.5). Other examples might include noting maximum and minimum values of system components, and proportional relationships among system components (ratio between A and B, or noting that A should go down as B goes up). The fact that some of our expectations represent the very hypotheses we want to test may seem trivially obvious. Theoretically this always should be the case. Unfortunately, all too often, we succumb to the temptation of leaving our expectations implicit and, hence, imprecise, only to find that the model is incapable of generating some important aspect of system behavior we need to answer our questions.

Quantitative model specification

Chapter outline

5.1 Overview of the process

The goal of Phase II of the simulation modeling process is to develop a quantitative model of the system-of-interest (Fig. 5.1). Using the conceptual model as a template for this quantitative development, we describe the rules governing the flow of materials in the model (the dynamics of the system) using mathematical equations. The first step is to select a general quantitative structure for the model. One useful mathematical format for quantifying the types of conceptual models we have developed in Chapter 4 consists of a set of difference equations, one for each state variable. From one time step to the next, the value of each state variable is calculated as the net difference between the material transfers that enter and leave the state variable. The rate of each material transfer is calculated based on information

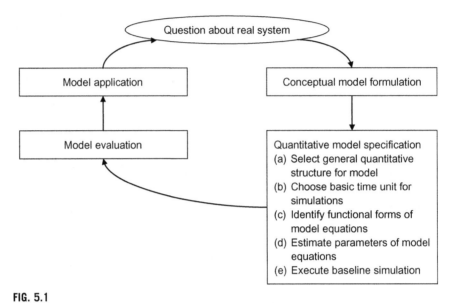

FIG. 5.1

Steps within Phase II of the simulation modeling process: quantitative model specification.

transferred to it from the appropriate constants, driving variables, auxiliary variables, and state variables.

Having selected the general quantitative structure, we next must develop the specific equations that collectively comprise the model. This consists of choosing the basic time unit for the solution of the model equations (e.g., 1 minute, 1 month, 1 year), the functional forms of model equations (e.g., linear, exponential, sigmoid, sinusoidal), and estimating the values of the parameters of the model equations. The best type of information we can use to develop the model equations is data from the real system. Available data often can be analyzed using standard statistical procedures to quantify the various aspects of the model. However, most commonly there will be some aspects of the model for which no data are available and for which we cannot collect new data at the present time. In such cases, we may be able to use information based on theoretical or generally applicable empirical relationships. Alternatively, we may be able to convert qualitative information, either from the technical literature or from "expert opinion," into a quantitative form. Finally, we may be able to use information generated from experimentation with the model itself to gain insight into the quantification of certain relationships.

The final steps in the quantitative specification of the model involve executing the baseline simulation and formally presenting model equations. The baseline simulation involves solving the model, or simulating the behavior of the system-of-interest, under a specific set of conditions that often represent the "normal" situation for the system, or are analogous to the "control" treatment in a designed experiment. The model equations are presented formally by listing them sequentially in a logical order that unambiguously describes how to solve the model.

5.2 **Select general quantitative structure for the model**

Theoretically, we should be able to represent the dynamics generated by the relationships among system components equally well in different mathematical formats. That is, the results of simulations should not depend on the particular mathematics (e.g., matrix algebra, differential equations, difference equations) or computer languages (e.g., FORTRAN, C++, Visual Basic) or computer programs (e.g., STELLA (isee systems, Inc.), VENSIM (Ventana Systems, Inc.) we use to represent them. In practice, however, this is not always the case. A discussion of the variety of different types of mathematical formats suitable for representing models of dynamic ecological systems is beyond the scope of this book (see Gilman and Hails, 1997; Jørgensen and Bendoricchio, 2001 for discussions of mathematical formats). However, one of the simplest, most flexible mathematical formats consists of a set of difference equations developed within a general compartment-model (box-and-arrow) structure. This compartment model structure is exactly the structure we have used thus far in this book to represent qualitatively our conceptual models and to do the arithmetic associated with some of the examples. The basic structural module is a state variable (Section 4.4.1) with a material transfer (Section 4.4.2) entering and/or leaving it. We "solve" the model, that is, we simulate system dynamics, by recalculating the value of each state variable each time step. We calculate the new value of each state variable by adding the value of each material transfer entering the state variable and subtracting the value of each material transfer leaving the state variable. For example, the equation we would solve each time step (Δt) for a state variable (SV) with one material transfer entering (MTE) and two material transfers leaving ($MTL1$ and $MTL2$) would be

$$SV_{t+1} = SV_t + (MTE_t - MTL1_t - MTL2_t) \times \Delta t \qquad (5.1)$$

Note that for difference equations, Δt always is equal to one. For example, Δt was equal to 1 minute, 1 day, 1 year, and 1 year in our bathtub (Section 3.6.1, Fig. 3.2), animal weight (Section 1.2.5, Fig. 1.1), nonrenewable natural resource (Section 3.6.2, Fig. 3.4), and population (Section 3.6.3, Fig. 3.6) models, respectively. Material transfers are calculated based on information transferred (Section 4.4.4) from other parts of the system, including information about state variables, constants (Section 4.4.5), driving variables (Section 4.4.6), and/or auxiliary variables (Section 4.4.7).

The general strategy is to write a set of equations that determine at selected points in time the value of each of the driving variables, auxiliary variables, material transfers, and state variables. Collectively, these equations are the quantitative model. To simulate behavior of the system-of-interest, we first specify the initial conditions of the system (at time zero), including the initial values of all of the state variables and the values of all of the constants. For each unit of simulated time, we then we "solve" the model by calculating the equations in the following sequence: (1) driving variable equations (if present), (2) auxiliary variable equations (if present), (3) material transfer equations, and (4) state variable equations (Fig. 5.2). We followed

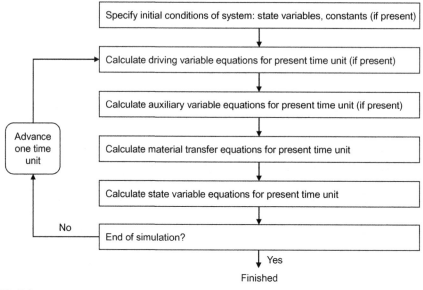

FIG. 5.2

Computing sequence for the model equations within the general compartment model structure.

exactly this procedure to calculate the dynamics our bathtub (Section 3.6.1, Eqs. 3.1–3.7), animal weight (Section 1.2.5, Eqs. 1.1–1.11), nonrenewable natural resource (Section 3.6.2, Eqs. 3.8–3.14), and population (Section 3.6.3, Eqs. 3.15–3.27) models.

5.3 Choose basic time unit for the simulations

After the general quantitative structure is chosen, the next step in quantitative specification of the model is to choose the basic time unit for simulations, which is the time interval (Δt) between iterative solutions of model equations. Our choice of the basic time unit depends on the level of temporal resolution needed for (1) addressing our questions and (2) appropriately representing temporal changes in the rates at which processes within the system-of-interest occur. Implicit in the choice of the basic time unit is the assumption that all of the rates within the system remain constant over any given Δt, which follows logically from the fact that we calculate each rate just once per Δt. Thus, for example, our choice of Δt equal to one year for the nonrenewable natural resource model connoted the assumptions that (1) the rate of resource decline could be adequately described by simulated annual changes in the amount of resource remaining, and (2) the resource extraction rate could be represented adequately as a simulated net annual rate. However, suppose for a moment we suspected that the extraction rate (measured as a proportion of the amount of resource remaining) varied markedly from month to month, as indicated in Table 5.1. In this

Table 5.1 Month-month variation in the rate at which a nonrenewable natural resource can be extracted from the earth.

Month	Year 1	Year 2	Year 3	Year 4	Year 5	Year 6	Year 7	Year 8	Year 9	Year 10
1	0.1	0.05	0.1	0.1	0.1	0.05	0.1	0.15	0.1	0.1
2	0.2	0.2	0.2	0.2	0.2	0.2	0.2	0.2	0.2	0.2
3	0.2	0.2	0.2	0.2	0.2	0.2	0.2	0.2	0.2	0.2
4	0.2	0.2	0.2	0.2	0.2	0.2	0.2	0.2	0.2	0.2
5	0.1	0.1	0.1	0.05	0.15	0.1	0.15	0.1	0.05	0.05
6	0.05	0.05	0.05	0.05	0.05	0.05	0.05	0.05	0.05	0.05
7	0.1	0.1	0.1	0.1	0.1	0.1	0.1	0.1	0.1	0.1
8	0.1	0.1	0.1	0.1	0.1	0.1	0.1	0.1	0.1	0.1
9	0.01	0.01	0.01	0.01	0.01	0.01	0.01	0.01	0.01	0.01
10	0.05	0.05	0.05	0.05	0.05	0.05	0.05	0.05	0.05	0.05
11	0.05	0.05	0.05	0.05	0.05	0.05	0.05	0.05	0.05	0.05
12	0.05	0.05	0.05	0.05	0.05	0.05	0.05	0.05	0.05	0.05
Average	0.1	0.1	0.1	0.1	0.1	0.1	0.1	0.1	0.1	0.1

Monthly extraction rates are measured as a proportion of the amount of the nonrenewable resource remaining, with the rates presented as the equivalent annual rates. The equivalent annual rates (AR) can be converted to the actual monthly rates (MR) used in the model as: $MR = 1 - (1 - AR)^{(1/12)}$. For example if $AR = 0.1$, $MR = 0.0087$.

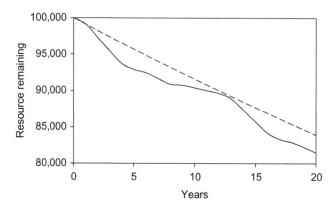

FIG. 5.3

Simulated declines in the amount of nonrenewable natural resource remaining assuming a constant rate of extraction of 10% per year *(dotted line)* versus the extraction rates in Table 5.1 *(solid line)*, which average 10% per year but vary from month to month.

case, we probably would want to shorten the Δt to 1 month in order to capture the effect of these monthly changes in extraction rates on rate of resource decline, even though our main interest still was focused on the net annual rate of decline. Fig. 5.3 shows the simulated declines in the amount of resource remaining assuming a constant rate of extraction of 10% per year versus the extraction rates in Table 5.1, which average 10% per year but vary from month to month. Whether or not the differences between these two projections are important would depend on the specific context of the problem at hand. Our point here is simply to emphasize that the importance of the choice of the basic time unit for simulation, with the concomitant assumption that all of the rates within the system remain constant over that time unit, should not be underestimated.

Note the basic time unit need not be confined to familiar units such as 1 day, 1 month, or 1 year. We might define Δt as 12h (night, day), 3 months (4 seasons per year), 365.25 days (length of the earth's orbit around the sun), or any other length that allows us to address our questions and represent temporal dynamics of the system adequately.

5.4 Identify functional forms of the model equations

The next step in quantitative specification of the model is choosing the functional forms of the model equations. We now specify whether the general forms of equations representing specific relationships within the model are linear, exponential, sigmoidal, or some other, perhaps more complicated, form. It is important to note here that we are referring to the functional forms of relationships between two, or among more than two, variables in the model. We are not referring to the resulting trends exhibited in a time series of values of the affected material transfers or state variables, such as we focused on in Chapter 2. For example, in Chapter 2, we presented a variety of what we referred to as building blocks, or modules, of system structure and function. We described the temporal trends in the value of the state variable(s) comprising each module as being, for example, linear or exponential or sigmoidal (Fig. 2.1). But the equations in the module generating linear growth or decline in the state variable did not have the form of an equation for a straight-line. The linear growth or decline in the state variable was generated simply by adding or subtracting a constant to the value of the state variable each time step. Likewise, the equations in the module generating exponential growth or decline in the state variable did not have an exponential form. The exponential growth or decline in the state variable was generated simply by multiplying the value of the state variable by a constant greater than one or less than one each time step. Nor did the equations in the module generating sigmoidal growth in the state variable have a sigmoidal (logistic) form. In that case, the sigmoidal growth in the state variable (SV) was generated by a functional relationship between the state variable and the auxiliary variable (Av) representing per capita rate of increase, which was used to calculate the material transfer (Mt) entering the state variable (Figs. 5.2–5.4). However, the functional form of the relationship was linear (Eq. 5.4). That is, per capita rate of increase decreased linearly with increasing values of the state variable:

$$SV_{t+1} = SV_t + (Mt) \times \Delta t \tag{5.2}$$

$$Mt = C4 \times Av \tag{5.3}$$

$$Av = (1 - Cc4 \times SV_t) \tag{5.4}$$

where in this case, $C4 = 0.1$ and $Cc4 = 0.01$.

So, to refocus, in this section, we are referring to the mathematical formulations that can be used to represent the relationships between two, or among more than two, variables in the model.

5.4.1 Information upon which to base choice of functional forms

Information upon which to base the choice of functional forms of model equations is of four general types: (1) quantitative data from direct observation or experimentation with the real system, either collected firsthand or from the literature; (2) information based on theoretical relationships or generally applicable empirical relationships that are appropriate for the situation being modeled; (3) qualitative information based on expert opinion or on the literature; and (4) information gained from experimenting with the model itself.

Quantitative data
We commonly think that the best information upon which to base the choice of functional forms comes from quantitative experimentation with, and/or quantitative monitoring of, the real system. This often is the case. However, we should keep in mind

that field data are limited by our technological abilities for measurement and subjectively biased by our perceptions of the system (Rykiel, 1996). The validity not only of the data but also of its interpretation ideally should be demonstrated convincingly. Given the relative inaccuracy and imprecision of ecological data, we should not automatically assume that field data provide a more accurate representation of the real system than other possible sources of information. Thus, it follows that we should not automatically rule out alternative sources of relevant information.

Theoretical relationships or generally applicable empirical relationships

In cases where data describing a particular process within our system-of-interest are not available, we may be able to rely on theoretical relationships or generally applicable empirical relationships to suggest the functional form of that process. For example, according to population growth theory, populations grow logistically (exhibit a sigmoidal growth form) in a limited environment. The differential equation for logistic population growth can be found in virtually all introductory ecology textbooks.

$$dN/dt = rN\left((K-N)/K\right) \tag{5.5}$$

where N is population size, K is the carrying capacity of the environment for the population, and r is the intrinsic rate of population increase. Eq. (5.5) embodies the negative feedback that slows the rate of population growth as population size increases, which, theoretically, is the result of increased intraspecific competition for limited resources. We have reformatted Eq. (5.5) as a compartment-model-based difference equations in Fig. 5.4.

An example of a generally applicable empirical relationship is the well-known relationship between body size (W) and metabolic rate (MR) ($MR = aW^b$, Kleiber, 1947), which has been parameterized successfully for a wide variety of animals. For example, French et al. (1976) estimated average daily metabolic rate (MR, in $\text{kcal}\,g^{-1}\,day^{-1}$) as a function of live body weight W, in g) for grassland rodents as:

$$MR = 2.297 W^{-0.50} \tag{5.6}$$

We have modified our animal weight model from Chapter 1 to illustrate use of Eq. (5.6) in a compartment model based difference equations in Fig. 5.5.

Qualitative information

In some cases, there may be no quantitative data, theoretical relationships, or generally applicable empirical relationships available upon which to base the choice of functional forms. In such situations, we may rely on qualitative information from the literature or on the opinions of experts in the field to establish assumptions upon which to base the choice of the functional forms of some of the model equations. For example, in our modified animal weight model (Fig. 5.5), we specified the functional form of the material transfer equation representing consumption simply as a constant proportion of current animal weight ($con = 0.25 \times W$). Assuming that no reliable

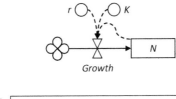

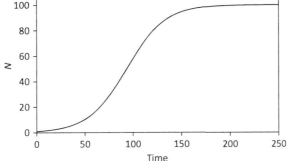

FIG. 5.4

A compartment model representing logistic population growth (Eq. 5.5 reformatted as a compartment model based on difference equations).

$$N_{t+1} = N_t + (growth)\,\Delta t$$

$$Growth = r \times N \times ((K-N)/K)$$

N is population size, K is the carrying capacity of the environment for the population, and r is the finite rate of population increase. Simulation results are based on $N_0 = 1$, $K = 100$, and $r = 0.05$.

quantitative estimates of the relationship between consumption and the weight of the animal were available, we might have based our choice of the rate constant representing that relationship on the opinion of a person with extensive experience in feeding such animals, even though written records concerning the rate of consumption never were maintained.

Quantifying qualitative information perhaps seems a much less rigorous procedure than analyzing data, and, indeed, most of us trained in the natural sciences feel much less comfortable quantifying simulation model equations in this manner. However, we would suggest that quantifying qualitative information is not necessarily a less rigorous procedure than data analysis, only less precise. In both cases, our goal is to use those techniques that allow us to interpret appropriately the available information to quantify important relationships within our system-of-interest. We almost always know more about our system-of-interest than we can substantiate quantitatively. Likewise, we almost surely will make a bigger mistake by excluding important system processes from our model because we lack the data to support their inclusion than we will by guessing (making a preliminary hypothesis about) how they might be quantified. Finally, we should keep in mind that

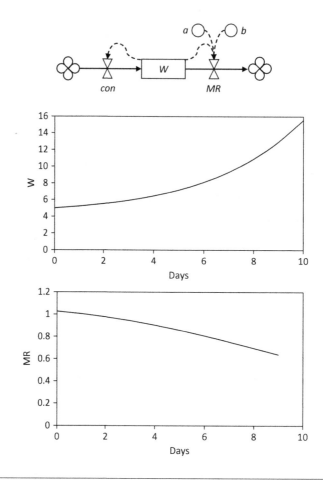

FIG. 5.5

Modified version of the animal weight model from Chapter 1 illustrating use of Eq. (5.6) ($MR = 2.297W^{-0.50}$) in a compartment model based on difference equations. Simulation results are based on $W_0 = 1$, $a = 2.297$, $b = -0.50$, and $con = 0.25 \times W$.

the numerical procedures we use to deal with the uncertainty in model parameters (e.g., sensitivity analysis, Section 6.5) work equally well regardless of whether our uncertainty originates from a lack of confidence in quantifying qualitative information, or from a lack of confidence in the accuracy of data.

Information from experimenting with the model

In some cases, there may be little known in even a qualitative sense upon which to base our choice of functional forms. In such situations, we may be able to gain insight into possible functional forms of a model equation by hypothesizing different functional forms and observing model behavior in response to each. Through such experimentation with the model, we may narrow the possible choices by excluding those functional forms that produce unreasonable model behavior. Returning to our modified animal weight model (Fig. 5.5), having specified the functional form of the material transfer equation representing consumption as a constant proportion of current animal weight ($con = 0.25 \times W$) based on expert opinion, we might want to explore other possibilities. For example, we might have hypothesized that the per capita consumption rate decreased linearly with increasing animal weight ($con = 0.25 - 0.0125 \times W$), and that the per capita consumption rate increased linearly with increasing animal weight ($con = 0.25 + 0.0125 \times W$). After examining the simulation results generated by each of these versions of the model, we might have decided that a linear decrease in per capita consumption rate was the most appropriate of the three function forms (Fig. 5.6). Obviously, the number of equations that can be specified in this manner within any single model necessarily is small. The more equations specified by this trial-and-error method, the higher the likelihood that we will obtain apparently reasonable model behavior purely by chance.

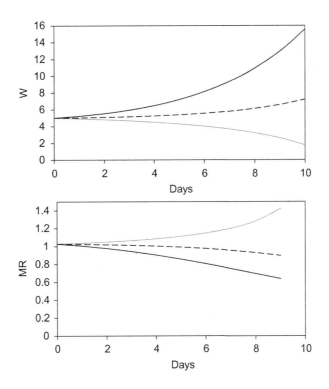

FIG. 5.6

Simulation results of alternative versions of the animal weight model from Chapter 1 illustrating the exploration of different functional forms of the material transfer equation relating consumption to current animal weight. The alternative versions assumed that the per capita consumption rate was a constant proportion of animal weight ($con = 0.25 \times W$), decreased linearly with increasing animal weight ($con = 0.25 - 0.0125 \times W$), and increased linearly with increasing animal weight ($con = 0.25 + 0.0125 \times W$).

5.4.2 Selecting types of equations to represent chosen functional forms

After we have chosen the functional form for a relationship in our model, we may be able to find a single equation that can be parameterized to represent the appropriate curve adequately. In our simple examples thus far, we have used linear, exponential, and logistical (sigmoidal) functional forms, all of which can be represented with a single equation. These equations, and some other useful equations for more complex functional relationships, are summarized in Fig. 5.7. Some logical procedures (dichotomous keys) for selecting appropriate equations to represent functional relationships are available in the applied mathematical literature (e.g., Spain, 1982, pp. 47 and 341).

Although delving too deeply into technical mathematical details would distract us from the main theme of this book, there are a couple of noteworthy considerations when choosing the types of equations with which we construct our models. Both deal with the extent to which the particular mathematical manipulations that constitute the model help us to think about our system-of-interest. The first consideration concerns the relative merit of a correlative versus an explanatory (or "mechanistic") model. We develop correlative models primarily to describe and summarize a set of relationships, without regard for the appropriate representation of processes or mechanisms that operate in the real system. The goal is prediction, not explanation. A model representing metabolic rate of an animal solely as a function of time of year might be an example. We develop explanatory models primarily to represent the internal dynamics of the system-of-interest with mathematical expressions that, for us, are interpretable in ecological (or other appropriate subject matter) terms. The goal is explanation through representation of the causal mechanisms underlying system behavior. A model representing the metabolic rate of an animal as a function of body mass, level of activity, reproductive status, and environmental temperature might be an example. Solely explanatory models and solely correlative models obviously form two idealized ends of a continuum, and classification relative to this dichotomy is based as much on the intent of the modeler as on the structure of the model. A model we view as explanatory at one level of detail we might view as correlative at a finer level of detail. We will return to this idea of correlative versus explanatory models when we consider the criteria for model evaluation in Chapter 6.

The second consideration concerns the relative merits of an analytical model versus a simulation model. This consideration is of more importance if our goal is to develop an explanatory model. Some mathematical models are solvable in "closed form," that is, they have a general analytical solution that applies to all of the situations the model can represent. Other mathematical models have no analytical solution, that is, the best mathematicians have not been able to solve them. These models must be solved step by step numerically, like the simulation models we are focusing on in this book. Conceptually, the choice between analytical and simulation models often involves deciding whether we would rather sacrifice

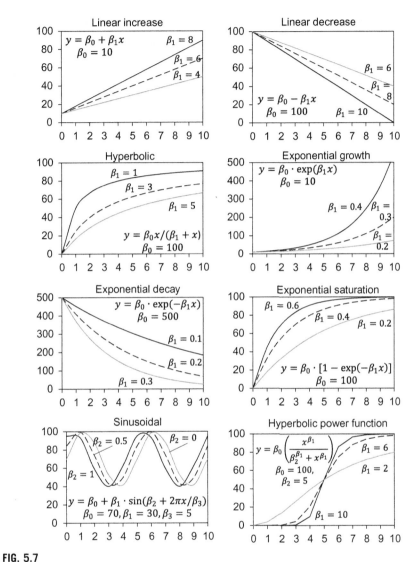

FIG. 5.7

Examples of some possible functional forms of the model equations.

ecological realism to obtain mathematical elegance (and the analytical "power" that goes with it), or we would rather sacrifice mathematical elegance to include explicitly more ecological realism. If the level of detail at which the system-of-interest must be represented to meet our objectives permits the use of an analytical model, then we should use an analytical model. However, if the appropriate level of detail requires a model that is too complex to be represented in an analytically solvable form, then we should use a simulation model. When developing explanatory ecological models, we most often will need to represent the system-of-interest in a manner too complex for analytical treatment. That is, we will need to use simulation models.

As an example of the two considerations discussed above, we might take another look at our model of logistic population growth, which we represented both as an analytically solvable model based on a differential equation (Eq. 5.5) and as a simulation model based on difference equations (Fig. 5.4). A first point to note is that there is no need to use a simulation model if, indeed, Eq. (5.5) is ecologically interpretable at an appropriate level of detail for our current interests. In this regard, r can be interpreted as the intrinsic rate of population increase, K can be interpreted as the carrying capacity of the environment for the population, and the term $((K-N)/K)$ can be interpreted as the density-dependent, negative effect of intraspecific competition on the rate of population growth. Thus, as population size approaches carrying capacity (as N approaches K), the rate of population growth approaches zero (the term $((K-N)/K)$ approaches zero, thus rN $((K-N)/K)$ approaches zero, and dN/dt approaches zero). The integral form of (analytical solution to) Eq. (5.5), which also can be found in virtually all introductory ecology textbooks, is

$$N_t = K/\left(1 + e^{b-rt}\right) \tag{5.7}$$

where b is a constant of integration defining the position of the curve relative to the origin. We note that it is difficult to estimate the value b in Eq. (5.7), thus we also provide the following, more detailed, form of the equation:

$$N_t = KN_0e^{rt}/\left[K + N_0\left(e^{rt} - 1\right)\right] \tag{5.8}$$

Using Eq. (5.8), we can calculate analytically the size of the population at any instant t, without the need to calculate numerically (via our simulation model) all of the preceding population sizes. For example, using Eq. (5.8), we can calculate that the population size at $t=100$ is 59.98596. To calculate the population size at $t=100$ using our simulation model (Fig. 5.4), we need to simulate (solve numerically), step by step, the new population size each year. The simulated population size at $t=100$ is 58.0571. The difference between the simulated population size and the population size calculated analytically is due to the implicit assumption associated with our choice of the one year as the basic time unit for simulation. Namely, that all of the rates within the system remain constant over any given Δt (see Section 5.3). With the analytical model, population size, as well as the rate of change in population size, are changing continually (instantaneously, in the calculus sense).

Thus, depending on our modeling objectives, simple, analytical models, such as the logistic equation model of population growth (Eq. 5.5), might be viewed as explanatory (mechanistic) models. Models do not need to be complex mathematically to be considered explanatory. But models do not need to be too complex mathematically before we lose the ability to solve them analytically, to find a general mathematical solution. For example, in our logistic equation model, consider the mathematical intractability that almost surely would arise from simply relaxing the assumption that carrying capacity, K, is constant. Suppose we were interested in representing explicitly the effects of changes in climatic conditions (C) and land use patterns (L) on K ($K = f(C, L)$). We could imagine that even a relatively simple representation of f might defy analytical solution, and would need to be solved numerically, via simulation.

5.5 **Estimate parameters of the model equations**

The information upon which to base the parameterization of model equations is of the same four general types used to choose the functional forms of model equations: (1) quantitative data, (2) information based on theoretical or generally applicable empirical relationships, (3) qualitative information, and (4) information gained from experimenting with the model itself. In fact, the choice of functional forms and the parameterization of model equations for any given model most commonly are based on the same information. Nonetheless, conceptually, we should view the choice of functional forms and the parameterization of model equations as two distinct steps because the former usually has more profound implications concerning the ecological interpretations of the model structure than does the latter.

As a simple generic example, consider a model consisting of a single state variable (V) which receives material from a source via a single material transfer (m). Suppose that the state variable represents some entity that is alive, an accumulation of bacterial cells, or individual animals, or plant biomass. Further suppose that our objective is to develop a mechanistic model that explains, or helps us explain, the process governing the material transfer. From the literature, we have obtained two different functional forms for the equation governing m: (1) $m = k1$ and (2) $m = k2V$, where $k1$ and $k2$ are constants. We also have obtained two different estimates of the value of $k1$ (0.1 and 0.2) and two different estimates of the value of $k2$ (0.1 and 0.2). Based on this information, we have developed four versions of our model using each of the four alternative representations of the process governing the material transfer (m): (1) $m = 0.1$, (2) $m = 0.2$, (3) $m = 0.1\,V$, (4) $m = 0.2\,V$. We then simulated the dynamics of V over a period of 10 time steps using each of these four model versions, assuming the initial value of V was equal to 10 (Fig. 5.8). Finally, suppose that we have field data from our system-of-interest suggesting that the value of V could increase from 1 to 3 in 10 time units.

Based on the information in the preceding paragraph, how should we go about deciding which is the most appropriate version of the model? The final values of V generated by versions 1 through 4 of the model were 2.0, 3.0, 2.6, and 6.2, respectively. With versions 1 and 2 of the model, V exhibited linear growth, hence, the growth rate of V was density-independent. With versions 3 and 4 of the model, V exhibited exponential growth, hence, the growth rate of V was density-dependent.

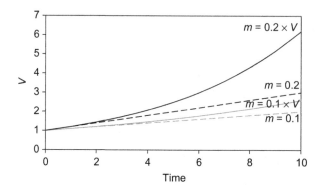

FIG. 5.8

Simulated dynamics of the state variable (V) using each of four alternative representations of the process governing the material transfer (m): (1) $m=0.1$, (2) $m=0.2$, (3) $m=0.1\,V$, (4) $m=0.2\,V$.

We would argue that since (1) the state variable represents some entity that is alive, and (2) our objective is to develop a mechanistic model that helps us explain the process governing the material transfer, version 3 of the model probably is the most appropriate. Growth of living entities is density-dependent in that the source of the new material (new cells, new individuals, new biomass) is the existing material. The fact that version 2 yielded a value of V after 10 time units that is closer to the expected value than that yielded by version 3 is not as important as the fact that the functional form of m in version 3 is density-independent whereas the functional form of m in version 2 is density-dependent. That is, the choice of functional forms has more profound implications concerning the ecological interpretations of model structure than does the choice of parameter values.

Specific techniques available to estimate parameters of model equations based on quantitative data are diverse, and a substantive discussion covering the spectrum of available techniques is beyond the scope of this book. Problems that arise within the framework of model parameter estimation are identical to those encountered in the analysis of data in general, and modelers are well advised to consult with professional statisticians during this step in the quantitative specification of the model. Nonetheless, the role of the various techniques of data analysis often employed during this step of model development should be clear. They serve as tools allowing us to interpret appropriately the quantitative information available about the important relationships within our system-of-interest.

5.5.1 Deterministic- versus stochastic-model parameterization

Another consideration regarding model parameterization is whether the model is deterministic or stochastic. Any model that contains one or more random variables is, by definition, stochastic (see Section 1.2.4). We use a stochastic model if we need to represent explicitly randomness, or uncertainty, in the dynamics of our system-of-interest in order to address our questions adequately. This uncertainty may come from randomness that we consider inherent within the system or from the uncertainty associated with our estimation of model parameters. Randomness can be associated with any model component, with the degree of variability that we assign to a component reflecting our level of uncertainty regarding the value, or values, the component should be given during a simulation. We will examine in detail the different levels, types, and sources of uncertainty encountered during the modeling process in Chapter 12.

Thus far in this chapter, we have been talking in terms of deterministic point estimates of model parameters, hence implying that we are dealing with deterministic models. We hasten to note here that variation is not synonymous with randomness. For example, the year-to-year variation of the extraction rate in our nonrenewable natural resource model (Table 5.1) was not random. Extraction rates were prescribed deterministically by the time series of values given to the driving variable representing extraction rate (extraction rate is a constant in Fig. 3.4). Although the same basic concepts just presented for deterministic models are equally applicable to stochastic models, some additional information is needed during the parameterization process in order to represent the inherent randomness associated with parameter values. The representation of a random variable in a stochastic model requires the specification of either a statistical distribution or a frequency distribution from which the values of the variable can be selected randomly, rather than the specification of a single-point estimate such as is required for deterministic models. As before, the information upon which to base the specification of the statistical or frequency distributions can be (1) quantitative data, (2) information based on theoretical relationships or generally applicable empirical relationships, (3) qualitative information, or (4) information gained from experimenting with the model itself.

To illustrate the differences in the information required to parameterize a deterministic versus a stochastic model, we might consider some possible alternative representations of the extraction rate in our nonrenewable natural resource model (Fig. 3.4). So far, we have represented the extraction rate as a constant (Fig. 3.5) and as a deterministic time series of different extraction rates that varied from year to year over the course of 10 years (Table 5.1, Fig. 5.3). Suppose we now decide to represent the extraction rate as varying randomly from year to year. Since we now will have a stochastic model, we will need to run more than one simulation in order to see how the random variability of the extraction rates affects the rates of decline in the resource. The results of each simulation will be different, purely by chance. We will address the question of how many simulations we should run and examine the results of several replicate stochastic (Monte Carlo) simulations in Section 5.6.2. But for now we will focus our attention on the parameterization process.

Suppose that we have data suggesting that we could represent the inherent year-to-year variability in the extraction rate as varying between 5% and 15%, with an equal probability of the occurrence of any extraction rate within this range each year. In this case, we might parameterize a statistical distribution to represent the extraction rate. More specifically, we could use a uniform distribution on the interval 0.05–0.15 (converting the percent extracted to the proportion extracted). Thus, for each year of simulated time, we would generate a random variate from a uniform distribution and assign the value of the random variate to the extraction rate. Or if we had data suggesting that we could represent the inherent year-to-year variability in the extraction rate as a normally distributed random variable with a mean of 10% and a standard deviation of 5%, we could generate a random variate from the corresponding normal distribution and assign the value of the random variate to the extraction rate. Since the normal distribution extends to plus and minus infinity, we would want to truncate the distribution at both ends. In this case, since we are dealing with percentages, which we have represented as proportions in the model, we would truncate the distribution at 0 and 1. In the vast majority of cases, we would truncate a normal distribution at zero at the lower end. If there is no logical upper bound to the variable we are dealing with, we would need to decide arbitrarily at what point the values become unreasonably large and truncate the distribution at that value.

Of course, there are numerous statistical distributions other than the uniform and the normal that find wide application in ecological modeling. The degree to which a data set resembles a particular statistical distribution, or the goodness of fit to that distribution, can be evaluated using standard statistical procedures (e.g., Johnson and Kotz, 1969). Algorithms for generating random variates from a variety of statistical distributions that we might wish to use in stochastic simulation models are readily available (e.g., Hastings and Peacock, 1975).

Quite obviously, not all of the variables that we might wish to represent as random variables in a simulation model exhibit a distribution that can be estimated well with a statistical distribution. In such cases, we could construct a cumulative frequency distribution based the frequency distribution of the available data, generate a random variate from the cumulative distribution, and assign the value of the random variate to the variable in question. For example, suppose that we had the field data on the annual extraction rates of our nonrenewable natural resource that are presented in Table 5.2. We could categorize these data in the form of a frequency distribution (Fig. 5.9A), and use this frequency distribution to construct a cumulative frequency distribution (Fig. 5.9B). The number and width of categories in such frequency distributions are arbitrary decisions based on the level of detail required of the random variable in question, given the objectives of the model. The height of each bar in the cumulative frequency distribution represents the proportion of observed extraction rates that is less than the value indicated beneath the bar. For example, the proportion of the observed extraction rates that were ≤ 0.1 is 0.1, the proportion of the observed extraction rates that were ≤ 0.2 is 0.11, the proportion of the observed extraction rates that were ≤ 0.3 is 0.19, and the proportion of the observed extraction rates that were ≤ 1.0 is 1 (Fig. 5.9B). To select an annual extraction rate randomly from the cumulative frequency distribution, we select a uniform random variate on the interval zero to one and note the category in the cumulative frequency distribution to which the value of the random variate corresponds. We then assign value of the extraction rate associated with that category in the

Table 5.2 Summary of field data on the monthly rates at which a nonrenewable natural resource has been extracted from the earth.

Extraction rate category	Number of months with indicated extraction rate
0–0.1	10
0.1–0.2	1
0.2–0.3	8
0.3–0.4	5
0.4–0.5	10
0.5–0.6	19
0.6–0.7	7
0.7–0.8	15
0.8–0.9	13
0.9–1.0	12

Extraction rates are measured as a proportion of the amount of the nonrenewable resource remaining, and are partitioned into the indicated categories.

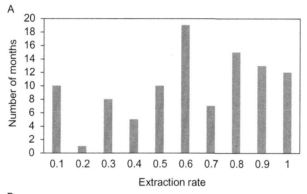

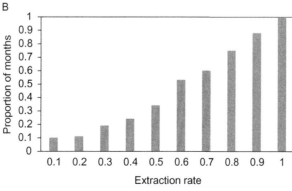

FIG. 5.9

Field data on the monthly extraction rates of the nonrenewable natural resource summarized in the form of (A) a frequency distribution and (B) a cumulative frequency distribution.

cumulative frequency distribution to the random variable representing the extraction rate in the model. For example, if the value of the uniform random variate is 0.55, we would assign the extraction rate a value of 0.7, if the value of the uniform random variate is 0.51, we would assign the extraction rate a value of 0.6, and if the value of the uniform random variate is 0.21, we would assign the extraction rate a value of 0.4 (Fig. 5.9B).

5.5.2 Statistical analyses within the context of simulation model parameterization

In addition to the use of statistical distributions to represent stochastic variables described in the previous section, we commonly employ statistical techniques to help us determine the appropriate quantitative representations for particular relationships in simulation models when data pertaining to those relationships are available. Often, regression equations resulting from the statistical analyses of the relationship between or among model components actually become a part of the simulation model. The statistical considerations that arise within the framework of model parameter estimation are identical to those encountered in analysis of data in general, and a detailed treatment of specific statistical procedures is beyond the scope of this book. Excellent statistical texts are readily available (e.g., Ott and Longnecker, 2001). For our purposes, the important point is simply that we use those statistical techniques that allow us to interpret appropriately the available data to quantify important relationships within our system-of-interest.

5.6 Execute baseline simulation

The baseline simulation represents the behavior of the system-of-interest, or the solution of the model, under the particular set of conditions that we wish to use as a benchmark or standard against which to compare the changes in system behavior resulting from other sets of conditions of interest to us. Note that prior to the formal running of the baseline simulation, we must verify that the computer code is correct (verify or "debug" the code). That is, we must make sure that the computer program is doing what we want it to do. For example, if we want the code to add 2 plus 2, the code must generate a result of 4. The baseline simulation is the end product of the second phase of the simulation modeling process just as the conceptual model diagram is the end product of the first phase. As we will see in Chapter 6, the process of model evaluation involves the close examination of the baseline simulation, and the comparison of the baseline simulation projections of various system attributes with the corresponding attributes observed in the real system. As we will see in Chapter 7, the process of applying the model to address our objectives involves the comparison of the baseline simulation results with the results of the simulations representing the management policies or environmental situations that we wish to examine.

Thus, the baseline simulation often is analogous to the "control" treatment in the design of a field or laboratory experiment. For example, in our nonrenewable natural resource model, we stated that current regulations limit the amount of the resource extracted annually to 10% of the estimated amount remaining (Section 3.6.2). We then proceeded to wonder how many more years the resource would last if we decrease the allowed annual extraction rate from 10% to 5% or 2.5% of the estimated amount remaining. Within this context, the experimental design for our simulated experiment would consist of a baseline (control) simulation assuming a 10% extraction rate, and two other simulations (treatments) assuming a 5% and a 2.5% extraction rate, respectively. However, just as in the design of a field or laboratory experiment, we do not always have a control treatment, in the statistical sense, in a simulated experiment. For example, in our bathtub example (Section 3.6.1), we observed that our bathtub fills up at different rates depending on how far we open the faucet, and we wondered how many minutes it would take for the bathtub to fill up with water if we open the faucet one turn versus two turns versus three turns. Within this context, our experimental design would not include a statistical control treatment. None of the three treatments is more "normal" or "benchmark" than the others, and we would take this into consideration when analyzing the simulation results. Nonetheless, for practical purposes during model evaluation, we almost always identify one set of "benchmark" conditions under which we execute the baseline simulation, which serves as a point of reference for our model evaluation procedures.

5.6.1 **Present the baseline model equations**

As part of this final step in the quantitative specification of the model, we formally present the model equations. We present them formally by listing them sequentially in a logical order that unambiguously describes how to solve the model. Such formal presentation facilitates the unambiguous description of the model structure to others. Although there are various formats that might be used, one effective manner in which to present the model equations is to use the format we have followed thus far to describe the models in our simple examples. We begin with a state variable equation and then sequentially define each term in the equation, either by providing another equation for the term or by presenting a specific numerical value if the term is a constant. (See, e.g., Eqs. 3.1–3.3, 3.8–3.10, and 3.15–3.21 in Chapter 3, which describe our bathtub, nonrenewable natural resource, and population models, respectively.) After defining all of the terms in the first state variable equation, we proceed to the next, ideally following a sequence that can be related easily to our conceptual model diagram. An alternative format is to present the model equations organized by the computing sequence of the equation types, following the order presented in Fig. 5.2. In this case, the initial conditions of the state variables and the constants of the model are presented first, followed by the driving variable equations, the auxiliary variable equations, the material transfer equations, and the state variable equations.

5.6.2 **Baseline simulations for stochastic models**

Whereas there is only one baseline simulation for deterministic models, the baseline simulation for stochastic models actually consists of a set of replicate stochastic (or Monte Carlo) simulations, each representing the baseline conditions.

The simulation results of each replicate will be unique due to the chance variation of the random variable, or random variables, in the model. Thus, for stochastic models, we must address the question of how many replicate stochastic simulations we should execute.

This question is directly analogous to the question of how many samples should we obtain in an experiment in the real world. As in the real world, there is no simple, universal answer to the sample size question. But, in general, the answer depends on (1) the inherent variability of items we are sampling, and (2) the magnitude of the difference between the different samples of items that we consider to be practically significant relative to our project objectives. We might rephrase the question more formally statistically (following Sokal and Rohlf, 1969, p. 246) as: "How large a sample must we obtain in order to show that a true difference of magnitude δ between alternative management policies or environmental situations is statistically significant at a significance level α with a probability P that the difference will be detected if it exists?" Then we could draw upon the sample size formula found in basic statistics textbooks to estimate quantitatively how large a sample we need, or how many replicate stochastic simulations we should execute (e.g., Ott and Longnecker, 2001, p. 221). We can estimate the variability of items we are sampling by running relatively many replicate stochastic simulations of the baseline conditions and calculating the variance of the model output, or outputs, in which we are interested. Although it is difficult to define "relatively many," as we increase the number of these initial simulations, the estimate of the variance of an item should reach a stable level. We must determine the magnitude of the differences between samples that we consider to be practically important independent of the model, based on our knowledge of what size differences are relevant in the real system, given our objectives.

The formula for calculating the number of samples needed to detect a given true difference between sample means, assuming that we have an estimate of variability within samples, is

$$n \geq 2\left(\frac{\sigma}{\delta}\right)^2 \left[t_{\alpha\gamma} + t_{2(1-P),\gamma}\right]^2 \tag{5.9}$$

(Sokal and Rohlf, 1969, p. 247)
where

$n =$ number of samples
$\sigma =$ true standard deviation, which we estimate as the square root of the estimated variance within samples
$\delta =$ smallest true difference that we desire to detect
$\gamma =$ degrees of freedom of the sample standard deviation with b groups of samples and n samples per group, or $\gamma = b(n-1)$
$\alpha =$ significance level
$P =$ desired probability that a difference will be found to be significant if it is as small as δ
$t_{\alpha\gamma}$ and $t_{2(1-P),\,\gamma} =$ values from a two-tailed t-table with γ degrees of freedom and corresponding to probabilities of α and $2(1-P)$, respectively

We cannot solve for n directly because we do not know γ, which is a function of n. Therefore, we (1) guess a value for n, (2) calculate γ, and (3) solve the equation for n. We then compare the calculated n with our guess of n. If the calculated n rounded to the nearest whole number is not equal to the guessed n, we adjust our guess for n accordingly and repeat the procedure.

As an example of estimating an appropriate number of replicate stochastic simulations to execute, we now return to the stochastic version of our nonrenewable natural resource model, which we described in Section 5.5.1. We will focus on the version in which we generated the random extraction rate using a normally distributed random variable with a mean of 10% and a standard deviation of 5%. We first address the two key sample size questions: (1) what is the inherent variability of items we are sampling, and (2) what is the magnitude of the difference between the different samples of items that we consider to be practically significant relative to our project objectives? We might rephrase these questions as: (1) what is the inherent variability of each of the model outputs in which we are interested, and (2) what is the magnitude of the difference among the different scenarios we will simulate during model application for each of the model outputs that we consider to be practically significant relative to our project objectives? Recall that the model output of most interest was the number of years that it would take to exhaust the nonrenewable natural resource, where we defined "exhaust" as decreased to less than 1 unit (Section 3.6.2).

We have estimated the inherent variability of the simulated number of years to exhaust the resource (E) by running 100, and then 500, replicate stochastic simulations of the baseline condition. That is, we represented the annual extraction rate (er) as a normally distributed random variable with a mean of 0.1 and a standard deviation of 0.5, with the distribution truncated at 0 and 1. The resulting estimates of the standard deviation (σ) of E were 5.35 and 5.54, respectively. We will use $\sigma = 5.54$. We will assume that the magnitude of the difference in E among the different scenarios that is of practical significance within the context of our modeling project is 2 years, thus, $\delta = 2$. We will use $\alpha = 0.05$ and $P = 0.80$. Since eventually, during model application (a topic that we will discuss in detail in Chapter 7), we want to compare three scenarios (three groups of samples, with mean annual extraction rates of 0.1, 0.05, and 0.025), $b = 3$. To calculate the degrees of freedom of the sample standard deviation (γ), we need to make an initial guess at the value of n that will satisfy inequality 9 (Eq. 5.9). We will guess that $n = 50$. Thus, $\gamma = 3\,(50 - 1) = 147$, $t_{\alpha\gamma} = t_{0.05,147} \approx t_{0.05,\infty} - 1.960$, and $t_{2(1-P),\ \gamma} = t_{0.40,147} \approx t_{0.40,\infty} = 0.842$. We looked up the values of $t_{0.05,\infty}$ and $t_{0.40,\infty}$ in a table of the critical values of the Student's t-distribution. Therefore, our initial estimate of the number of replicate stochastic simulations to execute becomes:

$$50 \geq 2\,(5.54/2)^2\,(1.960 + 0.842)^2 \tag{5.10}$$

or

$$50 \geq 60.2415 \tag{5.11}$$

Fifty is not greater than or equal to 60, so, to be precise, we could repeat the procedure with another guess for the value of n. However, note that for practical purposes, the improvement in our estimate of n gained by iteration often is unimportant. Usually, we arbitrarily set $\gamma = \infty$ and use the initial calculated value of n as our estimate of the required sample size, realizing that it is a slight underestimate. Therefore, we will assume that $n = 60$ is a good estimate of the number of replicate stochastic simulations we should execute.

If we examine the results of the 60 replicate stochastic baseline simulations, we note that, indeed, the trajectories of the decline in the nonrenewable resource are different in the different simulations (Fig. 5.10A), which leads to a distribution of values of E (Fig. 5.10B). The mean time to exhaust the resource is about 108 years, with a standard deviation of about 5 years.

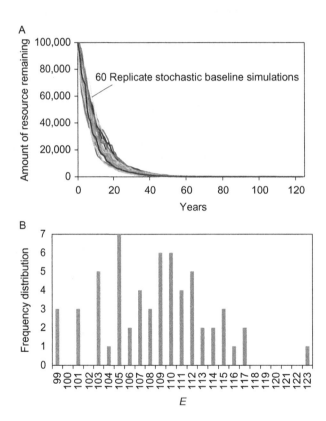

FIG. 5.10

(A) Simulated declines in the nonrenewable natural resource and (B) the resulting distribution of values (frequency distribution) of E (number of years to exhaust the resource), assuming that the annual extraction rate is represented as a normally distributed random variable with a mean of 0.1 and a standard deviation of 0.5, with the distribution truncated at 0 and 1. Results are based on 60 replicate stochastic baseline simulations.

Model evaluation

Chapter outline

6.1 Overview of the process

The goal of Phase III of the simulation modeling process is to evaluate the model in terms of its relative usefulness for a specific purpose. A model that is very useful for one purpose might be useless or, worse, misleading for another purpose. Most commonly, this process is referred to as "model validation," erroneously suggesting that a "valid" or "correct" model exists for any given system-of-interest. Holling (1978), among others, recognized this was an unfortunate choice of terms and suggested that we might better refer to the process of "model invalidation," making an analogy between the process of attempting to invalidate a model and the process of attempting to refute a hypothesis via the scientific method. This analogy has merit in that a model can be viewed as a collection of hypotheses that represents our current understanding of the structure and function of the system-of-interest. However, it tends to overemphasize the role of quantitative, often statistical, comparisons of the model predictions with selected observations from the real system as the sole criterion for evaluation.

Rykiel (1996) argues convincingly that different validation criteria are appropriate for different types of models and suggests that validation should mean simply that a model is acceptable for its intended use. Rykiel also emphasizes that no generally accepted standards currently exist for the validation of ecological models and provides an excellent discussion of the semantic and philosophical debate concerning validation, in which ecological modelers still are involved. We will reflect in more detail on the semantics, philosophical viewpoints, methodological approaches, procedures, and criteria involved in model "validation" in Chapter 10. Although the details of the debate may seem quite confusing, we believe that the basic ideas involved in evaluating the usefulness of a model are easy to understand. Thus, we prefer to refer simply to the process of "model evaluation" and to focus our attention on the examination of the various characteristics of a model that make it a potentially useful tool.

During model evaluation, we should examine a broad array of qualitative as well as quantitative aspects of model structure and behavior (Fig. 6.1). We begin by

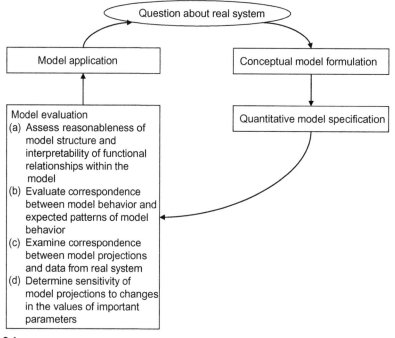

FIG. 6.1

Steps within Phase III of the simulation modeling process: model evaluation.

assessing reasonableness of model structure and interpretability of functional relationships within the model. Reasonableness and interpretability are defined relative to the ecological, economic, or other subject-matter context of the model. Next, we evaluate correspondence between model behavior and the expected patterns of model behavior that we described during conceptual-model formulation. We then examine more formally the correspondence between model predictions and real-system data, if available. These comparisons may or may not involve the use of statistical tests. Finally, we conduct a sensitivity analysis of the model, which usually consists of sequentially varying one model parameter (or set of parameters) at a time and monitoring the subsequent effects on important aspects of model behavior. By identifying those parameters that most affect model behavior, sensitivity analysis provides valuable insight into the functioning of the model and also suggests the level of confidence we should have in model projections. Relative importance of these various steps for any given model depends on the specific objectives of the modeling project.

6.2 Assess the reasonableness of the model structure and the interpretability of functional relationships within the model

This first step involves attempting to refute aspects of model structure and functional relationships within the model based on their lack of correspondence with the real system. Thus, this step receives particular emphasis with explanatory models and might be omitted for strictly correlative models. Explanatory models are those whose objectives require that relationships within the model be interpretable in ecological (or other subject matter) terms. Correlative models are those whose objectives require only that they project system dynamics adequately, regardless of the specific relationships that generate those dynamics. The procedure is exactly the same as scientific hypothesis testing. The hypotheses tested are our hypotheses about how functional relationships within the model work and about structural connections among individual parts of the model. Hypothesis tests are based on the best information available about the corresponding aspects of the real system, viewed at the appropriate level of detail. If any aspect of model structure or any functional relationship within the model can be shown to be an inadequate representation of the corresponding aspect of the real system, then that particular portion of the model is refuted. Criteria for failing hypothesis tests may be qualitative or quantitative, depending on objectives of the modeling project and the type of information available from the real system.

   This step is inherently subjective and, hence, somewhat difficult to describe precisely. For the three simple examples that we presented in Chapter 3, the bathtub model, the nonrenewable natural resource model, and the population model, it is relatively easy to assess the reasonableness of model structure based directly on the problem descriptions and some implicit assumptions which seem obvious. The bathtub needs to fill up, the nonrenewable natural resource needs to decrease, and the population size needs to change via births and deaths. It is difficult to imagine alternative model structures. We assumed no leaks in the bathtub, the nonrenewable natural resource, by definition, cannot be replenished, and we implicitly assumed no immigration to, or emigration from, the oceanic island population. (In Chapter 4, we relaxed these assumptions by adding (to the conceptual models) a leak to the bathtub, a replenishment of the natural resource via discovery of new deposits, and immigration to the population.)

For our present purposes of illustrating this step in model evaluation, we will refocus on our example dealing with the oceanic island population, but we will modify the statement of our interest, that is, we will modify the objective of the model. In Section 3.6.3, we stated our interest as follows: "We wonder how many years it will take for the population to recover to its pre-hurricane size." We will modify the statement to "We wonder what the effect of hurricane frequency has on ability of the population to recover to its pre-hurricane size." Thus, we now are interested not only in the rate of population recovery from the most recent hurricane, but also in the effects of subsequent hurricanes on population dynamics.

Assessing the reasonableness of our original model structure in view of our new interests, we probably would think that it would be more reasonable to represent explicitly both the occurrence of hurricanes and the effect of hurricanes on the death rate of the population. In our original model, both the occurrence of the recent hurricane and the effect of the hurricane on the death rate of the population were implicit in the initial condition of the population (Fig. 6.2A). One way to include hurricane frequency explicitly would be to represent it as a constant (Fig. 6.2B). Another way to include hurricane frequency explicitly would be to represent it as a driving variable (Fig. 6.2C). The effect of hurricanes on the death rate of the population might be represented as a density-independent or a density-dependent, additive or compensatory, agent of mortality. These choices, obviously, depend on our assumptions about the frequency and effect of future hurricanes within the broader context of our interests in the modeled population. Let's assume we think that the probability of a hurricane occurring during any given year is likely to be 10% over the next 100 years. Let's also assume we think that each hurricane is likely to kill an additional 40% of the population. That is, 40% in addition to the percentage of the population that die each year from nonhurricane-related causes.

Assessing the interpretability of functional relationships within our modified model, our original functional relationship between births and population size ($births = (1 - 0.00125 \times Pop) \times Pop$, see Section 3.6.3) is interpretable ecologically in terms of intraspecific competition and population growth in a limited environment. That is, in a limited environment, the per capita availability of resources decreases as population size increases (density-dependent negative feedback of population size on per capita birth and/or death rates), which results in a logistic (S-shaped) form of population growth. Our original functional relationship between deaths and population size in nonhurricane years ($deaths = 0.50 \times Pop$, see Section 3.6.3) is interpretable ecologically in terms of density-independent mortality. Likewise, our new functional relationship between hurricane-related deaths, and nonhurricane-related deaths and population

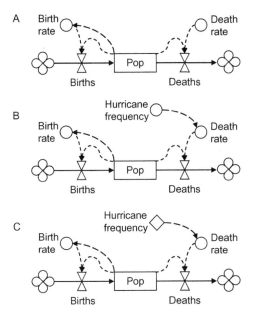

FIG. 6.2

Conceptual models representing (A) how fast an oceanic island animal population can recover to its prehurricane size following a large hurricane, (B) the effect of constant hurricane frequency on the ability of the population to recover to its prehurricane size, and (C) the effect of variable hurricane frequency on the ability of the population to recover to its pre-hurricane size.

size in hurricane years ($deaths = (0.50 + 0.40) \times Pop$) is interpretable ecologically in terms of hurricanes representing an additive, density-independent agent of mortality. Our new stochastic representation of the frequency of occurrence of hurricanes is interpretable in terms of the inherent uncertainty of hurricane occurrence.

Of course, implicit in these evaluations of the interpretability of functional relationships is the identification of the level of interpretability we deem necessary to meet project objectives. That is, based on project objectives, where along the scale from purely correlative to purely explanatory should the model we have developed be positioned. A model cannot be positioned along the correlative-explanatory scale in any absolute sense. We know from general systems theory that there always are less detailed, less explanatory representations of the system-of-interest, and more detailed, more explanatory representations. Thus, the same model might be viewed as correlative from one perspective and explanatory from another. Nonetheless, through the modeling process, we will have identified at least an approximate "benchmark" on the correlative-explanatory scale against which we should evaluate the interpretability of functional relationships in our model. For example, in the oceanic island population model, almost surely we would have thought that the representation of hurricane-related deaths as a function of the time of arrival (day of the year) of the hurricane and the direction from which it came (north, south, east, west) would have been inappropriate because it contained too many unnecessary details. On the other hand, almost surely, we would have thought that the representation of the occurrence of hurricanes strictly as a function of time, that is, as occurring at precisely 10-year intervals, would have been inappropriately simple.

6.3 Evaluate the correspondence between model behavior and the expected patterns of model behavior

During the second step in model evaluation, we compare model behavior to our a priori expectations, which we described during the final step in conceptual-model formulation. In these comparisons, we are looking for obvious impossible or implausible aspects of the baseline simulation, such as negative values for state variables or material transfers that by definition must be positive, or impossibly high or low values for state variables or material transfers. In addition to examining the baseline simulation for unreasonable behavior, we might also examine model behavior over a wide range of input conditions (wide range of values for driving variables) to attempt to expose additional inadequacies in the model. It may seem strange to mention such obvious shortcomings in model behavior, but models that initially show gross inconsistencies with our expectations based on general knowledge about the real system are very common. At this point, these gross inconsistencies may result from a fundamental misconception about the nature of relationships within the system-of-interest, in which case changes in the conceptual model are required. Or, they may result from erroneous a priori expectations, which we only now identify as erroneous based on a better understanding of system processes gained from the model itself.

For example, related to the original version of our oceanic island population model described in Chapter 3, we noted that, given the negative feedback of population size on population birth rate, we would expect logistic (S-shaped) population growth toward carrying capacity (Section 3.6.4). And, in fact, model behavior exhibited the expected pattern. The simulated population grew from its relatively low post-hurricane size (10 individuals) to its prehurricane size of 400 animals per km^2 in approximately 19 years (Fig. 3.7). We may not have expected that the population would reach carrying capacity in exactly 19 years, but as long as we did not expect the population to reach carrying capacity in a much shorter (e.g., 5 years) or longer (e.g., 50 years) period of time, we probably would have concluded that model behavior corresponded with our expectations reasonably well. Thus, we would have evaluated the correspondence between model behavior and the expected patterns of model behavior as satisfactory.

Related to the modified version of our oceanic island population model described in this chapter, with the addition of probabilistically reoccurring hurricanes, we no longer would expect the "smooth" approach toward carrying capacity depicted in

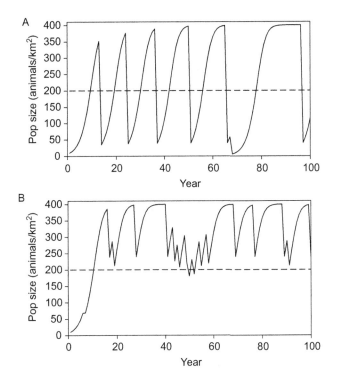

FIG. 6.3

(A) Expected pattern of population dynamics for the modified version of the oceanic island population model, and (B) simulated pattern of population dynamics resulting from a typical replicate stochastic (Monte Carlo) simulation.

Fig. 3.7. Rather, we might expect a repeated, "jagged," approach toward carrying capacity, with population size being reduced to somewhere near 10 individuals immediately after a hurricane and then increasing logistically (in an S-shaped form) toward carrying capacity until the next hurricane occurred (Fig. 6.3A). However, if we use the modified version of our oceanic island population model to simulate population dynamics over the next 100 years, a typical replicate stochastic (Monte Carlo) simulation yields a population pattern such as that depicted in Fig. 6.3B. The equations comprising the modified model are the same as the equations comprising the original model (see Eqs. 3.15–3.21 in Chapter 3), except that the annual death rate in Eq. (3.21) is increased from 0.50 to 0.90 (0.50+0.40) in years when a hurricane occurs, and hurricanes are generated randomly each year with a 10% probability of occurrence. One numerical method to generate hurricanes with a 10% probability of occurrence is to generate a uniform random variate on the interval from 0 to 1.0 (virtually all computers are equipped with software that can do this), if the value of the random variate is less than 0.10 a hurricane occurs (and death rate is increased accordingly), if the random variate is more than 0.10 no hurricane occurs (and death rate is not increased). Refocusing our attention on the simulated pattern of population dynamics in Fig. 6.3B, we notice that population size does not decrease anywhere near 10 individuals after a hurricane. Not even when hurricanes occur, by chance, in a series of consecutive years, as occurred in the simulation depicted in Fig. 6.3B. Thus, we evaluate the correspondence between model behavior and the expected patterns of model behavior as unsatisfactory, and we are obliged to reconcile the difference between the two patterns. Either the model is flawed, our expectation was flawed, or both.

In this case, let's suppose that after reexamining the relationship between the occurrence of hurricanes and births in the population, we decide that births are, indeed, affected by hurricanes. Let's suppose that we decide that there are no births during a hurricane year. Perhaps hurricanes usually occur during the time when young are being born and none survive. Since our time step is one year, effectively, it would be as if no births had occurred. That is, there is no recruitment to the population. Thus, we would modify our conceptual model by adding an information transfer from the driving variable representing hurricane frequency to the auxiliary variable represent birth rate (Fig. 6.4A). We also would modify our functional relationship between births and population size such that no births occur during years in

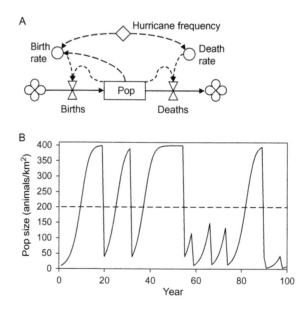

FIG. 6.4

Conceptual model representing (A) the effect of variable hurricane frequency, affecting both births and deaths in the population, on the ability of the population to recover to its prehurricane size, and (B) simulated pattern of population dynamics resulting from a typical replicate stochastic (Monte Carlo) simulation.

which there is a hurricane (if *hurricane frequency* = 1, *births* = 0). During nonhurricane years, we still would use our original functional relationship between births and population size (if *hurricane frequency* = 0, *births* = $(1 - 0.00125 \times Pop) \times Pop$). The simulated pattern of population dynamics generated by the modified model now includes decreases in population size the year following hurricanes to levels ranging between approximately 50 animals per km^2 and less than 5 animals per km^2 (Fig. 6.4B). If a hurricane occurs when the population size is near carrying capacity (400 animals per km^2), population size is reduced to about 50 animals per km^2. If a hurricane occurs when the population size is relative low (less than, say, 150 animals per km^2), population size may be reduced to levels around the expected level of 10 animals per km^2. Of course, our expectation of a population size of 10 animals per km^2 the year following a hurricane is based on information that did not include the prehurricane size of the population (see original problem description in Chapter 3, Section 3.6.3). Thus, assuming that it might have been relatively low, we probably would evaluate the correspondence between the modified model behavior and the expected patterns of model behavior as satisfactory.

When a model no longer exhibits obviously implausible behavior, we turn our attention to a closer examination of model components: state, driving, and auxiliary variables; constants; and material transfer rates. General dynamics of each component should be reasonable in terms of the timing of maximum and minimum values, the relative amplitude and periodicity of fluctuations, and its relationship to dynamics of other components. Both state variables and material transfers should vary in the right direction and by the right order of magnitude in response to changes in the values of driving variables. Inadequacies detected as a result of this closer examination of model components still may be caused by a fundamental misconception about the nature of relationships within the system-of-interest, but at this stage, it is likely that inadequacies are caused by inclusion of erroneous parameter estimates or perhaps by inclusion of incorrect functional forms of equations.

6.3.1 Tuning the model

At this point, assuming that no gross inadequacies in model behavior remain, it may be possible to further improve correspondence with expected patterns of model behavior by adjusting some parameter values or by altering functional forms of some equations. This general procedure often is referred to as "tuning the model" or model calibration. It is a legitimate activity during this step of model evaluation, but is an activity that must be confined strictly in terms of (1) choosing parameters or functional forms to be altered, (2) determining when to end tuning activities, and (3) limiting the number of parameters or functional forms tuned.

Choosing parameters or functional forms to be altered

Discretion must be used in choosing the parameters or functional forms to be altered and in choosing the types of alterations to be made. Alterations will be arbitrary because no new information external to the model is being drawn on to make them, although the nature of differences between model behavior and expected patterns of behavior may suggest specific alterations. But we must alter only those parameters or functional forms that we have specified in a tentative fashion, and the type of alteration made must not result in a parameter value or functional form that could have been refuted in the previous model-evaluation step.

For example, suppose that we have estimated two parameters in a given model. For one parameter, we have strong evidence that the value must be 8. For the other parameter, we have information suggesting only that the value probably lies between 10 and 20; perhaps we have chosen a value of 15 for inclusion in the model. Because we have strong evidence supporting our estimate of the first parameter, we should not "tune" or alter that parameter. But because our current estimate of the second parameter (15) was chosen over other possible estimates between 10 and 20 based on little or no information, we might legitimately tune or alter this parameter within the range of 10–20 in order to improve correspondence with expected patterns of model behavior.

Likewise, suppose that we have specified functional forms of two material transfer equations within a given model. For one transfer, we have strong evidence that the functional form is linear. For the other transfer, we have information suggesting only that the curve relating rate of transfer (Y) to the variable governing the transfer (X) passes through points ($X1$, $Y1$) and ($X2$, $Y2$). Perhaps we have chosen a linear relationship passing through these two points for inclusion in the model. Because we have strong evidence supporting our choice of a linear functional form for the first transfer, it should not be altered. However, because our current functional form of the equation representing the second transfer was chosen over other possible functional forms that pass through ($X1$, $Y1$) and ($X2$, $Y2$) based on little or no information, we might legitimately tune this transfer by specifying another functional form for the equation. For example, we might see if an exponential functional form of the equation improves correspondence with expected patterns of model behavior.

Determining when to end tuning activities

All tuning of the model must be completed before proceeding to the next model-evaluation step, which is the examination of the correspondence between model predictions and data from the real system. If statistical comparisons are made between model predictions and real-system data, then clearly tuning must be completed before statistical tests are performed. If comparisons between model predictions and real-system data are nonstatistical, tuning still must be completed before the final round of "formal" quantitative comparisons are performed.

Limiting number of parameters or functional forms tuned

We must limit the number of parameter values or functional forms of equations that we tune in any given model. Although there is no set limit, obviously, as the number of components that are tuned increases, so does the number of possible combinations of alterations and, hence, the possibility that any improvement in model behavior will result solely from a fortuitous combination of alterations.

In practice, it is more common to tune parameter values than functional forms of equations. This is understandable because the choice of functional forms usually has more profound implications concerning ecological interpretations of model structure than does the choice of parameter values. We are likely to have based our original choice of general functional forms on a relatively sound understanding of the nature of relationships in the real system, whereas we are less likely to have based our original choice of specific parameter values on an equally sound basis.

If the model fails to pass this second evaluation step, we should reconsider steps 3 and 4 of the quantitative-model specification, identifying the functional forms of the model equations and estimating the parameters of the model equations, and perhaps even reconsider step 4 of the conceptual model formulation, identifying the relationships among components that are of interest.

At this point in model development, there is an interesting philosophical consideration with regard to whether we are tuning an existing model or developing a new model after having completely rejected the old model. Perhaps the most practical distinction is based on the degree of our dissatisfaction with model behavior. If our dissatisfaction is relatively minor, we may tune the model as just described without use of new information external to the model. If our dissatisfaction is major, we may decide that the model is fatally flawed and begin developing a new model by drawing on new information about the system-of-interest or by changing basic assumptions about the structure or function of the system.

A numerical example of tuning

Before concluding this section, we want to present a numerical example illustrating some of our cautionary points about model tuning activities. We will use our animal weight model from Chapter 1 in which weight changed daily based on the net difference between consumption and respiration. Consumption (g day^{-1}) was a function of animal weight (g), and respiration (g day^{-1}) was a function of animal weight and environmental temperature (C) (see Eqs. 1.1–1.5 in Section 1.2.5 and Fig. 1.1). (Note that, in Chapter 5, we modified our animal weight model by changing the functional form of the material transfer equation representing consumption (see Eq. 5.6 in Section 5.4.1 and Figs. 5.5 and 5.6). However, for our present purposes, we will use the version of the model from Chapter 1.) In Chapter 1, we simulated weight change over a 2-day period, assuming that initial weight was 100 g and that temperature was 20°C on day 1 and 19°C on day 2. For the present example, we will simulate weight change over a 30-day period during which daily temperature decreases linearly from 20°C to 15°C ($temp_t = 20 - 0.167 \times t$, where t represents day of the experiment, day 0, day 1, ... day 29). Simulated changes in animal weight over the 30-day period are presented in Fig. 6.5A.

Let's suppose, for the sake of this example, that we had no reliable estimates available for either the coefficient in the consumption equation ($con = 0.05 \times W_t$) nor the coefficient in the respiration equation ($res = 0.0025 \times temp_t \times W_t$), but that we did have reliable estimates suggesting that an initially 100-g animal would weigh about

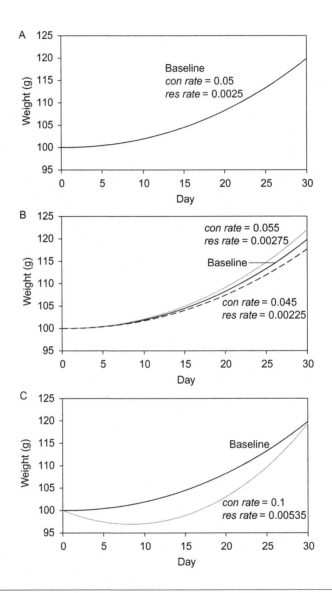

FIG. 6.5

Animal weight simulated over a 30-day period by various versions of the animal weight model from Chapter 1 using different pairs of coefficient values in the consumption and respiration equations. (A) Results of the original version with coefficient values in the consumption and respiration equations of 0.05 and 0.0025, respectively. (B) Results of the original version compared with results of two alternative versions with slightly different coefficient values. (C) Results of the original version compared with results of a version with much higher coefficient values.

120 g after 30 days if exposed to the declining temperature conditions being simulated. That is, suppose that the coefficient values of 0.05 and 0.0025 have resulted from our model tuning activities, which had the goal of obtaining a simulated animal weight close to 120 g at the end of the 30-day simulation. One could imagine that we might have discovered different pairs of coefficient values that would have yielded a simulated animal weight close to 120 g at the end of the 30-day simulation. For example, either raising each of the pair of coefficient values slightly (to 0.055 and 0.00275, respectively), or lowering each of the pair of coefficient values slightly (to 0.045 and 0.00225, respectively), yields a simulated animal weight close to 120 g (122 and 118 g, respectively) (Fig. 6.5B). But note that raising each of the pair of coefficient values by a relatively large amount (to 0.1 and 0.00535, respectively), also yields a simulated animal weight close to 120 g (119 g) (Fig. 6.5C). In fact, if our model tuning activities were conducted solely as a numerical exercise, without reference to the ecological or physiological implications of the coefficient values, the number of such pairs of values that we could discover would depend primarily on the amount of time and energy we were willing to devote to the tuning activity. And it follows that we also could discover pairs of values that would yield an animal weight closer and closer to 120 (.000…) g. But we would have no reason to choose one pair of coefficient values over another pair.

Thus, although tuning is a legitimate activity during this step of model evaluation, it is an activity that must be constrained, above all, by common sense with regard to (1) choosing parameters or functional forms to be altered, (2) determining when to end tuning activities, and (3) limiting the number of parameters or functional forms tuned. The simple example in the preceding paragraph illustrates that the decision to tune simultaneously two parameters that have opposite effects on the system component being monitored may be ill-advised. The rate of change in the value of a state variable tells us nothing about the rates of flow of material into and out of the state variable. With the relatively high values for both consumption and respiration coefficients, animal weight after 30 days was approximately the same as it was with relatively low values for both coefficients (Fig. 6.5). However, the manner in which the animal's weight changed over the 30-day period was noticeably different. With the high values, weight no longer exhibited a continual increase, but, rather, decreased during the first few days (Fig. 6.5C). Also, the rates of consumption and respiration were approximately double those obtained using the lower coefficient values (Fig. 6.6). Almost surely we would have enough basic knowledge about the animal's physiology to identify the consumption and respiration levels, and the resulting weight dynamics, generated by either the relatively low or the very high coefficient values as being more reasonable. However, if we decided the relatively low values generated more reasonable results, it probably would be difficult to choose between the two sets of low values. In fact, given the similar manner in which the animal's weight changed over the 30-day period, if the difference between 122 and 118 g was of no practical significance within the context of our modeling objectives, arguably, we should have ended our tuning activities sooner.

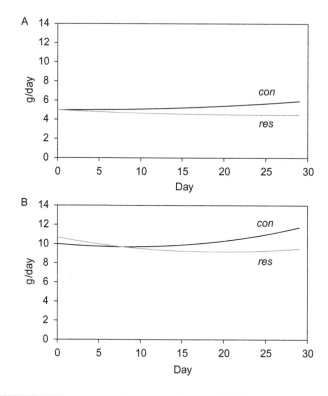

FIG. 6.6

Rates of consumption and respiration simulated over a 30-day period by various versions of the animal weight model from Chapter 1 using different pairs of coefficient values in the consumption and respiration equations. (A) Results of the original version with coefficient values in the consumption and respiration equations of 0.05 and 0.0025, respectively. (Results of two alternative versions with slightly different coefficient values (not shown) were essentially the same as the original version.) (B) Results of the original version compared with results of a version with much higher coefficient values.

That is, our "fine tuning" has tempted us into evaluating differences in animal weight based on unwarranted precision. Finally, this discussion of cautionary points illustrated by tuning two parameters in a very simple model should make it clear that the number of parameters tuned should be kept to a minimum.

6.4 Examine the correspondence between model projections and the data from the real system

The third step in model evaluation focuses more specifically on the correspondence between model projections and real-system data. This step receives particular emphasis with correlative models and theoretically might be omitted for strictly explanatory models, although in practice we almost always are interested to some extent in a model's ability to project system dynamics. Strictly speaking, data from the real system that are used in model evaluation must be independent of data used to develop the model. If the same data used in the quantitative-model specification are used in model evaluation, we can hardly expect to reject the model, because we would have already examined and failed to reject any aspects of the model that were quantified based on those data. The situation is analogous to quantifying a regression model using a given data set, concluding that we will tentatively accept the regression model because it has an r value of 0.90 and then "evaluating" the model simply by observing that it does indeed project values in that given data set well. The appropriate evaluation for the regression model is to use the model, as parameterized based on the first data set, to predict values of the dependent variable in a new data set based on values of the independent variables in that new data set. Likewise, for this step in simulation model evaluation to be meaningful, model projections must be compared to real-system data that were not used directly to quantify model equations.

Related to the need to use independent data to evaluate a simulation model, it is important to realize that we use different types of data in different ways during simulation model development. Commonly, data naturally sort themselves into three general categories: those used to (1) evaluate model behavior, (2) quantify driving variables, and (3) quantify auxiliary and material transfer equations (and associated constants). Data we use to evaluate the model often are time series of values of state variables. Data we use to quantify driving variables often are time series of values over the same historical period for which we have the time series of state variable values. Data we use to quantify auxiliary and material transfer equations are of various types, including time series data, but the distinguishing feature is that these data are analyzed (manipulated) in some appropriate way and the results of these analyses (manipulations) are used to write model equations.

Thus, ideally, we would have a group of data we could analyze to quantify each of the auxiliary and material transfer equations in the model, and time series data over the same time period for each state variable and each driving variable. We then could compare simulated to historical system behavior by "driving" or "forcing" the model with the time series of driving variable values and comparing the resulting time series of simulated values for each state variable, which were generated by the model equations, to the time series of data for each state variable. We also could compare any single simulated value to its historical counterpart.

For example, refocusing on our animal weight model, suppose that we had data sets from an experiment in which we had recorded daily environmental temperatures ($temp_t$, °C), rates of animal respiration (res_t, $g\,day^{-1}$), and the resulting animal weights (W_t, g) over a 30-day period (Table 6.1). We could parameterize the driving variable ($temp_t$) as a function of day of the experiment (t) via linear regression as: $temp_t = 20 - 0.167 \times t$. Likewise, we could parameterize the material transfer equation via linear regression as: $res_t = 0.0025 \times temp_t \times W_t$. We then could simulate animal weight change over a 30-day period (with $con_t = 0.05 \times W_t$) and compare the simulated daily animal weights to the observed daily animal weights (Fig. 6.7).

Table 6.1 Daily environmental temperatures ($temp_t$), rates of animal respiration (res_t), and animal weights ($obs\ W_t$) measured daily over a 30-day period, as well as simulated weights ($sim\ W_t$) produced by the animal weight model using the time series of driving variable values contained in $temp_t$.

Day of experiment (t)	Temp$_t$ (°C)	res$_t$ (g day^{-1})	obs W$_t$ (g)	(sim W$_t$) (g)
0	20.000	5.0000	100.0000	100.0000
1	19.833	4.9582	101.0970	100.0000
2	19.666	4.9186	103.8750	100.0418
3	19.499	4.8809	100.1650	100.1253
4	19.332	4.8451	101.6520	100.2507
5	19.165	4.8113	99.5953	100.4181
6	18.998	4.7793	98.7829	100.6277
7	18.831	4.7492	97.9351	100.8798
8	18.664	4.7208	100.7120	101.1746
9	18.497	4.6942	98.7899	101.5126
10	18.330	4.6693	104.0550	101.8940
11	18.163	4.6461	104.0700	102.3194
12	17.996	4.6245	100.9960	102.7893
13	17.829	4.6045	98.9237	103.3043
14	17.662	4.5862	104.1700	103.8650
15	17.495	4.5693	102.4510	104.4720
16	17.328	4.5541	108.5470	105.1263
17	17.161	4.5403	110.9310	105.8285
18	16.994	4.5280	1088160	106.5797
19	16.827	4.5172	109.6670	107.3806
20	16.660	4.5079	111.1480	108.2324
21	16.493	4.5000	114.5700	109.1361
22	16.326	4.4934	106.2050	110.0930
23	16.159	4.4883	107.3560	111.1042
24	15.992	4.4846	107.5210	112.1711
25	15.825	4.4822	116.0530	113.2950
26	15.658	4.4812	111.7290	114.4775
27	15.491	4.4816	118.1780	115.7202
28	15.324	4.4832	115.2900	117.0247
29	15.157	4.4862	121.8610	118.3927
30	14.990		118.0740	119.8261

(As you may have noticed, these are exactly the equations for *temp*, *res*, and *con* that we used in the animal weight model for our numerical example of tuning in Section 6.3.1. Thus, the simulated weight curve in Fig. 6.7 is identical to the weight curve in Fig. 6.5A.)

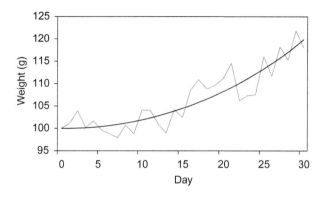

FIG. 6.7

Comparison of animal weights simulated over a 30-day period of linearly decreasing
environmental temperatures by the animal weight model with animal weights observed in an
experiment over a 30-day period under the same linearly decreasing environmental
temperatures (Table 6.1).

The manner in which we examine the correspondence between model projections
and data from the real system depends on the specific objectives of our modeling
project, the type of model we have developed, and the type of real-system data available. If we are interested in comparing values projected with a deterministic model
to nonreplicated data from the real system, we simply compare the simulated and
observed values, nonstatistically, and assess their relative difference or similarity
within the context of our problem. For example, since our animal weight model is
deterministic, assuming that we have only the nonreplicated observed weights
reported in Table 6.1 with which to evaluate the simulated weight changes, we simply would compare the simulated versus observed trajectories of weight change and
the simulated versus observed final weights at the end of the 30-day period. Whether
we would judge the correspondence between the two trajectories or the correspondence between the two final weights to be more important would depend on the
specific objectives of the modeling project. Likewise, whether we would judge
the differences between simulated versus observed weights in Fig. 6.7 to be of practical significance would depend on the specific objectives of the modeling project.

If we have a stochastic model and/or replication in data from the real system, our
comparisons of simulated versus real-system data might include statistical tests of
significance. We already should have determined the number of replicate stochastic
simulations we need to run at the time we executed the baseline simulations
(see Chapter 5, Section 5.6.2). An additional point worthy of mention here is that
time series of values projected by a simulation model are auto-correlated. This results
from the fact that the state of the system at $t+1$ depends explicitly on the state of the
system at time t. This auto-correlation among the time series of values generated
for any given system component during a single simulation violates the assumption
of independence of observations required by many parametric statistical tests.

We provide some examples of simple statistical tests below. For more detailed considerations of specific statistical tests, one should consult a basic statistics textbook (e.g., Ott and Longnecker, 2001). In any event, the same restrictions that apply to use of statistical tests on real-system data also apply to model projections (simulated data).

6.4.1 Comparisons of single value predictions

Single-value model predictions of interest often are summary statistics from an entire simulation, although they can be any single prediction from a simulation. For example, in our animal weight model, the weight of the animal on day 30 of simulated time and the total consumption by the animal over the 30-day period (the sum of all of the daily material transfer values, con_t) both are single-value predictions.

If we are interested in comparing a single-value prediction from a deterministic model to nonreplicated data from the real system, our comparison is nonstatistical (Table 6.2). We must base our comparison solely on the practical significance of the difference between model prediction and the real-system estimate.

If we are interested in comparing a single-value prediction from a stochastic model to replicated data from the real system, we can use a two-sample t-test to test the hypothesis that the mean of model predictions equals the mean of real-system data (Table 6.2). For example, suppose that we convert our animal weight model into a stochastic model by representing environmental temperature as a stochastic variable. For purposes of this example, we now will choose each daily temperature from a normal distribution with a mean equal to the temperature predicted by the regression equation representing temperature as a function of day of the experiment ($temp_t = 20 - 0.167 \times t$, where t represents day of the experiment) and a variance equal to 2.0 (arbitrarily chosen for this example). We will truncate the normal distribution on the lower end at 13°C and on the upper end at 22°C (also arbitrarily chosen for this example, the normal distribution, of course, extends from $-\infty$ to $+\infty$). Suppose that we decide that we want to be able to detect a true difference between simulated and observed animal weights of 3 g (δ) at a statistical significance level of $\alpha = 0{:}05$ with probability $P = 0.80$ that the difference will be detected if it exists. If we follow the procedure for estimating the required sample size described in Chapter 5 (Section 5.6.2), we will find that we need a sample size of $n = 14$. Thus, we will run 14 replicate stochastic (Monte Carlo) simulations of our animal weight model and record the weight of an initially 100-g animal on day 30 of simulated time for each. Assuming that we use a different set of random numbers in each replicate simulation to generate the time series of values of the environmental temperature driving variable, the 14 predictions of animal weight on day 30 form a set of independent, identically distributed random variates. If we also have 14 independent replicate observations of weight on day 30 of initially 100 g animals from the real system, we can compare the 14 replicate model predictions to the 14 replicate estimates from the real system using a t-test, as shown in Table 6.3.

Table 6.2 Summary of appropriate quantitative comparisons of model predictions to real-system data and associated procedures, depending on (1) the type of model predictions, (2) the type of model, and (3) the type of data available from the real system.

Type of predictions	Type of model	Type of real-system data	Appropriate comparison	Appropriate procedure		
				Statistical		Nonstatistical
				Hypothesis tested	Test statistic	
Single value	Stochastic	Replicated	Compare mean of model predictions (μ_{MP}) to mean of real-system data (μ_{RSD})	$\mu_{MP} = \mu_{RSD}$	$t = (\overline{X}_{MP} - \overline{X}_{RSD}) / \sqrt{(S^2_{MP} + S^2_{RSD})/n}$ See Table 6.3, for example, calculations and symbol definitions	
		Nonreplicated	Compare mean of model predictions to single real-system datum (which represents μ_{RSD})	$\mu_{MP} = \mu_{RSD}$	$t = (\overline{X}_{MP} - X_{RSD}) / (S_{MP}/\sqrt{n})$ See Table 6.4, for example, calculations and symbol definitions	
	Deterministic	Replicated	Compare single model predictions (which represents μ_{MP}) to mean of real-system data	$\mu_{MP} = \mu_{RSD}$	$t = (X_{MP} - \overline{X}_{RSD}) / (S_{RSD}/\sqrt{n})$ See Table 6.4, for example, calculations and symbol definitions	
		Nonreplicated	Compare single-model prediction to single real-system datum			X
Time-series	Stochastic	Replicated	Compare mean curve predicted by model to mean curve based on real-system data with regard to timing of maximum and minimum values and amplitude and periodicity of fluctuations		Statistical procedures may be possible to apply in specific cases but are beyond the scope of this book	
		Nonreplicated	Compare mean curve predicted by model to single curve based on real-system data		Statistical procedures may be possible to apply in specific cases but are beyond the scope of this book	
	Deterministic	Replicated	Compare single curve predicted by model to mean curve based on real-system data		Statistical procedures may be possible to apply in specific cases but are beyond the scope of this book	
		Nonreplicated	Compare single curve predicted by model to single curve based on real-system data			X

Table 6.3 Example comparing 14 replicate predictions of weight of an initially 100-g animal on day 30 of simulated time from the animal weight model to 14 replicate estimates of the weights on day 30 of initially 100-g animals from the real system using a two-sample t-test.

Model predictions (X_{MP}) of weight (g) on day 30	Real-system estimates (X_{RSD}) of weight (g) on day 30
122.3	177.5
122.6	116.0
120.6	122.5
123.4	117.1
121.1	126.6
120.2	116.9
120.4	118.0
114.8	119.6
119.1	121.5
119.4	118.8
119.3	114.9
118.4	121.2
119.1	117.4
117.4	120.1

Degrees of freedom, df = 26.
Test statistic, t = 0.703.
Significance level, $\alpha > 0.05$.

There are other statistical tests that also can be used to compare model predictions to real-system data when we have replication in both. If the values we wish to compare are not normally distributed (normality is an assumption of the t-test), we might use a nonparametric procedure such as a Wilcoxon signed-rank test (Snedecor and Cochran, 1967, p. 130) or a Kolmogorov-Smirnov test (Sokal and Rohlf, 1969, p. 571) to compare the frequency distributions of predicted and observed values. However, the t-test is quite robust to departures from normality and can be used in most cases.

If we are interested in comparing a single-value prediction from a stochastic model to nonreplicated data from the real system, or a single-value prediction from a deterministic model to replicated data from the real system, we can use a modified two-sample t-test to test the hypothesis that the mean of model predictions (or the mean of real system data) equals the single real-system datum (or the single deterministic-model prediction) (Table 6.2). We assume that the single datum is a sample from a population, but because the population is represented by a single variate, the sample does not contribute to the degrees of freedom or to the within-group variance calculated as part of the t-test. An example comparing the 14 replicate model predictions of weight of an initially 100-g animal on day 30 of simulated time

Table 6.4 Example comparing 14 replicate predictions of weight of an initially 100-g animal on day 30 of simulated time from the animal weight model to a single estimate of the weight on day 30 of an initially 100-g animal from the real system using a modified two-sample *t*-test.

Model predictions (X_{MP}) of weight (g) on day 30	Real-system estimate (X_{RSD}) of weight (g) on day 30
122.3	122.5
122.6	
120.6	
123.4	
121.1	
120.2	
120.4	
114.8	
119.1	
119.4	
119.3	
118.4	
119.1	
117.4	

Degrees of freedom, df = 13.
Test statistic, t = − 1.148.
Significance level, α > 0.05.

used in Table 6.3 to a single real-system estimate of weight on day 30 of an initially 100-g animal is shown in Table 6.4. The procedure for comparing a single model prediction to replicated real-system estimates is directly analogous.

6.4.2 Comparisons of time series predictions

Comparisons of time-series model predictions to time-series data from the real system follow the same general scheme just presented for comparisons of single-value predictions and data with regard to stochastic versus deterministic models and replicated versus nonreplicated real-system data (Table 6.2). The question concerning model evaluation is whether or not the curve based on the time-series of model predictions resembles the curve based on the real-system time series closely enough. Criteria defining "closely enough" will vary depending on the model objectives, but, in general, they will be more specific quantitative versions of the general criteria used to evaluate the correspondence between model behavior and the expected patterns of model behavior. Time-series predictions of interest often consist of sequential values of state variables or material transfers over the course of an entire simulation. For example, the 30 sequential values of daily animal weights and the 30 sequential values of daily consumption from our animal weight model both comprise time-series predictions. We might be interested in quantitatively examining the timing of maximum and minimum values and/or the amplitude and periodicity of fluctuations within these time series.

Suppose that the evaluation of our deterministic animal weight model includes the comparison of 30 sequential daily predictions of animal weight to sequential daily estimates of animal weight based on non-replicated real-system data. If the curves representing these two time series are as shown in Fig. 6.8A, we see that although the general shapes of the curves are similar, the model underestimates daily weights by 0.5 g to almost 8 g. Upon closer examination, we also note that the real-system data suggest a decrease in rate of growth at heavier weights, which is not reflected in model predictions, particularly during the last 8 days when the real-system data suggest a 5.5% increase in weight, whereas the model predicts a 9.5% increase. Whether these quantitative differences are sufficient grounds for refuting the model depends, once again, on the modeling objectives.

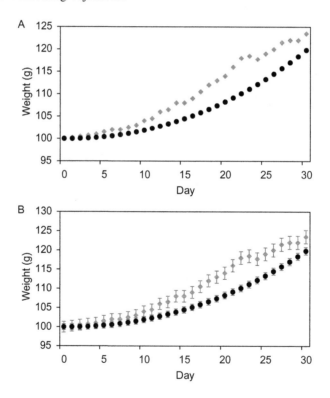

FIG. 6.8

Comparisons of the weight of an animal (A) based on deterministic-model predictions and nonreplicated real-system data, and (B) based on stochastic-model predictions and replicated real-system data. Horizontal bars represent ±1 standard error of the mean.

The use of statistical tests to compare time-series model predictions to time-series data from the real system is more complicated than is the case for single-value predictions and data. The complications arise because of auto-correlation between values within each time series, because often variances increase from the beginning to the end of simulations, particularly for state variables, and because for many of the situations of interest to us we have relatively few (less than 50) values within each time series. The exploration of these problems and the presentation of possible ways to overcome them in specific situations are beyond the scope of this book. One useful procedure when examining replicated time-series data is to represent the variance (or standard error) around each mean value in the two time series that we wish to compare, as illustrated for our animal weight model in Fig. 6.8B. In Fig. 6.8B, we note that the variation in the real-system data on daily weights is noticeably larger than the variation in the model predictions of daily weights. We also might examine the relative distance between the mean observed values and the predicted values compared to the widths of the bars representing the variability within the observed and the predicted values (the standard errors of the mean in Fig. 6.8B). The general interpretation is that if the bars are broadly overlapping, such as occurs for days 1 through 5, there is no "significant" difference between the observed and the predicted values. If the bars do not overlap (days 6 through 30), the difference may well be "significant." Note that this procedure does not imply a formal statistical comparison of the two curves. We could compare the observed and the predicted values statistically for any single day on the curves, for example, the final day, by treating the problem as a single-value comparison. Thus, we could use a t-test to compare the observed and the predicted animal weights on day 10, for example. However, having conducted this single test, we should not repeat the procedure for other days because our data are not independent from day to day. In fact, they are auto-correlated. The weight predicted on day 10 depends on the weights predicted on the preceding days and affects the weights predicted on the subsequent days.

6.4.3 Practical versus statistical significance

We should consider for a moment the distinction between practical significance and statistical significance within the framework of model evaluation. Theoretically, there are four possible combinations of practical and statistical significance that we might encounter when comparing model predictions to real-system data statistically. (We can, of course, consider model predictions to be significantly (importantly) different from real-system observations without performing a statistical test. Although, to avoid confusion, we probably would not use the term "significantly different.") Differences may be (1) both statistically significant and of practical significance to the modeler and/or resource manager, (2) both statistically nonsignificant and of no practical significance, (3) statistically significant but of no practical significance, or (4) statistically nonsignificant but of practical significance (Table 6.5).

Ideally, we would encounter only the first two of these cases, in which the practical significance and the statistical significance are synonymous. If we have designed our evaluation tests well, if we have selected the sample sizes capable

Table 6.5 Interpretations of the four possible combinations of practical and statistical significance that might be encountered when comparing model predictions to real-system data.

| Statistically | Practically | |
	Significant	Nonsignificant
Significant	Modeler/manager considers differences important, rejects model as not useful; statistical design of evaluation appropriate	Modeler/manager considers differences unimportant, fails to reject model as not useful; statistical design of evaluation overdesigned
Nonsignificant	Modeler/manager provided with no additional information from statistics; statistical design of evaluation underdesigned	Modeler/manager considers differences unimportant, fails to reject model as not useful; statistical design of evaluation appropriate

of showing that a true difference of a practically significant magnitude between the model predictions and the real-system data is statistically significant at the desired significance level (α) and probability level (P) (as described in the previous chapter), then this will be the case. The interpretation of such results is clear. If the differences are practically/statistically significant, we conclude that the model is not useful in meeting the objectives that we stated during the conceptual-model formulation. If the differences are practically/statistically nonsignificant, we conclude that the model passes this step in the evaluation procedure (Table 6.5).

However, if we have overdesigned or underdesigned our evaluation experiment, we may encounter cases in which there is a discrepancy between the practical and the statistical significance. If our sample size is too large (overdesign), we may have enough statistical power to declare relatively small differences between the model predictions and the real-system data, which are of no practical significance, statistically significant at the chosen α and P levels. But the interpretation of such results also is clear. We do not refute the model, because we have not detected practically significant differences (Table 6.5). Statistical significance alone is not relevant and we have wasted resources by overdesigning the experiment. If our sample size is too small (underdesign), we may not have enough statistical power to declare even practically significant differences statistically significant. Such results are not interpretable, or at least the statistical test provides no useful information. In fact, because we can identify statistical tests that have insufficient power to detect practically significant differences a priori, such statistical tests should not be conducted (Table 6.5).

As an example of the overdesigned and the underdesigned experiments, consider the following situation. Suppose that we want to detect a true difference of 10 units between the mean of the model predictions and the mean of the real-system data for a particular variable of interest. The level of practical significance is 10 units. Further, suppose that based on the initial estimates of the variability of our variable of interest, we determine that the appropriate sample size needed to detect a true difference of 10 units at the desired α and P levels is 50. If we run 500 replicate stochastic simulations, assuming that we also have 500 replicate observations available from the real system, our experiment is overdesigned. Not only can we detect a difference of 10 units at the desired α and P levels, we can detect smaller differences if they truly exist. We might declare a difference of one unit statistically significant. But such a small difference has no practical significance for us. We care only that the difference is less than 10 units and we could have obtained this information from 50 samples. If we run five replicate stochastic simulations, assuming that we also have five replicate observations available from the real system, our experiment is underdesigned. Even if our predicted mean differs from the observed mean by more than 10 units, we still lack the confidence that the difference is not due purely to chance.

6.4.4 Quantitative versus qualitative model evaluation

We also should consider for a moment the relative merits of qualitative as opposed to quantitative methods of model evaluation. The tendency is to think of quantitative methods, particularly those involving statistical tests, as being more rigorous and definitive than qualitative methods. We should use quantitative methods of evaluation whenever appropriate. However, we also should keep in mind that objective quantitative methods ultimately rest on subjective judgments relative to the appropriateness of their use in the evaluation procedure. Even statistical hypothesis tests rely on subjective judgment to determine the appropriate significance levels for a given problem. Thus, quantitative methods are not inherently better than qualitative methods, and both should be viewed within a framework that emphasizes the evaluation of the ability of the model to meet the project objectives.

6.5 Determine the sensitivity of model projections to changes in the values of important parameters

The fourth step in model evaluation is to perform a sensitivity analysis on the model. The objective is to determine the degree of response, or sensitivity, of model behavior to changes in the various model components (Smith, 1973; Steinhorst, 1979). The basic procedure is to alter the value of one parameter at a time by a specified amount throughout an entire simulation and observe the subsequent effect on the selected aspects of model behavior. This procedure may be modified to include varying sets of parameters at the same time and also may include varying the functional forms of the relationships within the model. By identifying those parameters or relationships to which model behavior is most responsive, or sensitive, sensitivity analysis provides an indication of the relative accuracy with which each parameter or relationship ideally should be estimated. This information is useful in determining the degree of confidence that should be placed in the model based on the confidence with which the most influential parameters or relationships have been estimated, and also is useful in establishing future research priorities.

We design the sensitivity analysis by specifying the parameters or relationships altered, the manner in which they are altered, and the specific aspects of model behavior monitored. Details of the design depend on the objectives of the modeling project, but commonly include altering selected constants and auxiliary variables (usually those that represent functional relationships) over a range of values representing our degree of uncertainty in their true values, and monitoring selected state variables, material transfers, and auxiliary variables (usually those that convert model "output" into a particularly useful form). A possibility with relatively small models is to alter each parameter and each combination of parameters over many values within the selected ranges. However, simulating all possible combinations of even relatively few levels of relatively few parameters quickly becomes an overwhelming task. Most commonly, we restrict our sensitivity analysis to relatively few parameters of particular interest. We decide if we will use statistical tests to compare model sensitivity to the changes in one parameter with model sensitivity to changes in another parameter based on the same general considerations described for comparison of the model projections to the real-system data (Section 6.4).

To provide a numerical example of sensitivity analysis, we will refocus on the most recent version of our island population model, which we described in Section 6.3. Recall that we had decided that there are no births, or at least no young that survive, during a hurricane year, and we had modified both our conceptual model (Fig. 6.4A) and our functional relationship between births and population size (if *hurricane frequency* = 1, *births* = 0) to reflect the effect of hurricanes on births. During nonhurricane years, we still used our original functional relationship between births and population size (if *hurricane frequency* = 0, *births* = (1 − 0.00125 × *Pop*) × *Pop*). Hurricanes were generated randomly each year with a 10% probability of occurrence. We will focus our sensitivity analysis on the coefficient (0.00125) representing the slope of this density-dependent relationship between birth rate and population size.

The ecological interpretation of changing the value of this coefficient is that it changes our representation of the rate at which intraspecific competition for limited resources increases as the population grows, which affects the carrying capacity of the island for the population. Since the death rate in nonhurricane years is 0.5, the population is in equilibrium (at carrying capacity in the absence of hurricanes) when the birth rate is equal to 0.5. If we make the slope steeper (make the value of the slope coefficient more negative), the birth rate will be equal to 0.5 at a smaller population size (carrying capacity will be reduced). If we make the slope less steep (make the value of the slope coefficient less negative), the birth rate will be equal to 0.5 at a larger population size (carrying capacity will be increased) (Fig. 6.9A–C). In this example, as the slope coefficient (S) increased from 0.00025 to 0.00125 to 0.00225, the carrying capacity in the absence of hurricanes decreased from 2000 animals per km^2 to 400 animals per km^2 to 222 animals per km^2 (Fig. 6.9D).

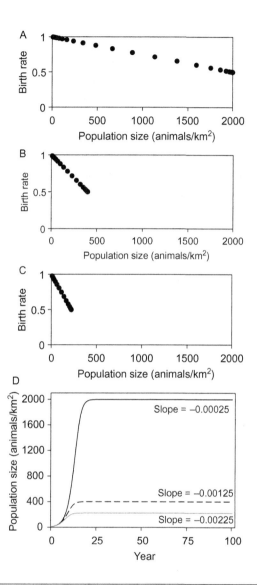

FIG. 6.9

Graphical illustrations of the relationship between the slope coefficient (S) in the functional relationship between births and population size in the oceanic island population model (*births* = (1 − *S* × *Pop*) × *Pop*) and carrying capacity of the island for the population in the absence of hurricanes (*K*). Birth rate versus population size with *S* equal to (A) 0.00025, (B) 0.00125, and (C) 0.00225, and (D) population growth associated with the indicated value of *S*.

We will suppose that the objective of our sensitivity analysis is to assess the effect that our uncertainty regarding the representation of the density-dependent relationship between birth rate and population size has on our estimate of the amount of time that the population will consist of less than 50 animals per km². That is, the amount of time that the population will be at risk of chance extinction due to low population size following a hurricane. Our sensitivity analysis will consist of running 100 replicate stochastic simulations, using each of 9 values for the slope coefficient (*S*) (ranging from 0.00025 to 0.00225). (For the present example, we will not repeat the calculations for determining the appropriate sample size, which we described in Chapter 5, Section 5.6.2.) Each simulation will last 100 years, and we will assume a 10% probability of a hurricane occurring each year. During each simulation, we will monitor the number of years during which the population size is less than 50 animals per km². We have summarized the results of the sensitivity analysis in Fig. 6.10. The cumulative number of years (during the 100 replicate stochastic simulations, each 100 years in length) in which the population size was less than 50 individuals increased from 3564 to 5465 as *S* increased from 0.00025 to 0.00225. Thus, our uncertainty regarding the representation of the density-dependent relationship between birth rate and population size has a noticeable (we will assume practically significant, see Section 6.4.3) effect on our estimate of the amount of time that the population will consist of less than 50 individuals. We will make use of this information from our sensitivity analysis in Chapter 7, when we return to our oceanic island population model to consider the experimental design for our model application simulations.

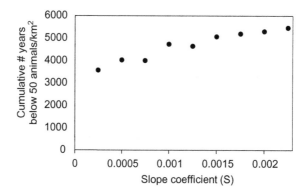

FIG. 6.10

Summary of sensitivity analysis results for the oceanic island population model. The graph depicts the sensitivity of the cumulative number of years (during 100 replicate stochastic simulations, each 100 years in length) in which the population size was less than 50 animals per km² to the indicated changes in the slope coefficient (*S*) in the functional relationship between births and population size (*births* = (1 − *S* × *Pop*) × *Pop*).

6.5.1 **Interpreting sensitivity analysis within a model evaluation framework**

The interpretation of sensitivity analysis results differs somewhat from the interpretation of the results obtained in the earlier steps of model evaluation. In the earlier steps, we attempt to refute the model as being useless. Through sensitivity analysis, we attempt to evaluate more clearly our level of confidence in the ability of the model to address our questions. If we have failed to refute the model during the first three steps of the evaluation procedure, we must consider the model irrefutable based on current knowledge about the real system. However, this does not necessarily mean that we have great confidence in the model's ability to answer our questions. Model behavior may be sensitive to changes in some parameters or functional forms of equations that we have estimated based on inadequate information. Our lack of confidence in estimates of these influential parameters or functional forms translates into a lack of confidence in the model projections.

One way to deal with this lack of confidence in the model projections due to the uncertainty in parameter estimates is to carry out our model application simulations (see Chapter 7) in parallel with different versions of the model, each with a different estimate for the given parameter, or parameters. Another option for dealing with this type of uncertainty is to represent the "uncertain" parameters as random variables, with the degree of their variability reflecting our level of uncertainty in their estimates. Neither option is inherently better than the other; our choice depends, as always, on our specific objectives. In Chapter 7, we will return to our oceanic island population model to illustrate these options.

Model application

7

Chapter outline

7.1 Overview of the process

The goal of the final phase of the simulation modeling process is to meet the objectives that were identified at the beginning of the modeling project (Fig. 7.1). Most often, we wish to use the model to simulate system dynamics under alternative management strategies or environmental situations. The general scheme for model application exactly follows the steps involved in addressing a question through experimentation in the real world. We first develop and execute the experimental design for simulations, just as we would develop and execute a field or laboratory experiment. Next, we analyze and interpret the simulation results, again, using the same analytical tools and interpretive schemes that we would use with the results from the field or the laboratory. Finally, we communicate our results to the appropriate audience, often using the same media that we would use to report the results of field and laboratory studies.

7.2 Develop and execute experimental design for simulations

The same principles that apply to the design of experiments conducted in the real world (Cochran and Cox, 1957) apply to experiments conducted, or simulated, on the computer. Our a priori expectations, which we described during the final phase of

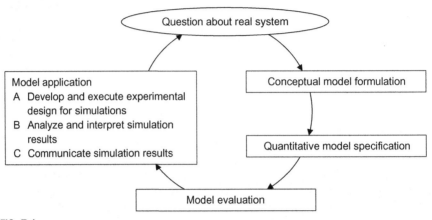

FIG. 7.1

Steps within Phase IV of the simulation modeling process: model application.

conceptual model formulation, should provide a good guide for the experimental design since some of these expectations represent the hypotheses that we want to test. We must avoid the temptation to abandon a well-planned experimental design in favor of the apparent expediency of a "shotgun" approach, which is made possible by the tremendous computing capabilities of modern computers. The ability to generate voluminous results does not preclude the need for a systematic, logical approach to the problem.

If our model is stochastic, as part of the experimental design, we also must specify the number of replicate simulations of each experimental treatment we will run. The question of how many replicate simulations of each alternative to run is directly analogous to the real-world question of how large a sample is needed. We discussed a method for determining the appropriate sample size in Chapter 5 (Section 5.6.2) when considering stochastic baseline simulations. In practice, the number of replicate simulations needed for comparisons of alternatives usually is established when running the baseline simulations during the last step of quantitative model specification.

As a numerical example of developing and executing an experimental design for simulations, we will return to our stochastic island population model from Chapter 6. We will design and execute a set of simulated experiments in which we assess the effect of an increased frequency of hurricanes on the ability of the population to sustain itself. Recall that the results of our sensitivity analysis in Chapter 6 (Section 6.5) indicated that our estimate of the amount of time that the population would consist of less than 50 animals per km^2, that is, the amount of time that the population would be at risk of chance extinction due to low population size following a hurricane, was sensitive to (affected by) our uncertainty regarding the representation of the density-dependent relationship between birth rate and population size. In our sensitivity analysis, we had simulated the effects of hurricanes using each of nine different representations of the density-dependent relationship between birth rate and population size, that is, using each of nine different values of the slope coefficient (S) (Fig. 6.10). In all of these simulations, hurricanes were generated randomly each year with a 10% probability of occurrence.

Given our uncertainty regarding the representation of the density-dependent relationship between birth rate and population size, we will carry out our model application simulations in parallel with different versions of the model, each with a different value of the slope coefficient (S). Thus, our experimental design to assess the effect of an increased frequency of hurricanes on the ability of the population to sustain itself will consist of simulating 36 scenarios. Using each of the nine values of S that we examined in our sensitivity analysis (0.00025, 0.0005, 0.00075, 0.001, 0.00125, 0.0015, 0.00175, 0.002, and 0.00225), we will simulate annual probabilities of occurrence of hurricanes of 10%, 12%, 15%, and 20% ($9 \times 4 = 36$ scenarios). As in Chapter 6, we will run 100 replicate stochastic simulations of each of these scenarios, and, during each simulation, we will monitor the cumulative number of years during which the population size is less than 50 animals per km^2. The results of our model application simulations are summarized in Fig. 7.2. We will set these results aside for a moment, but we will return to them in the next section when we consider the analysis and interpretation of simulation results.

Before concluding this section, we want to comment briefly on the other option for dealing with parametric uncertainty that we mentioned in Chapter 6 (Section 6.5.1). That is, to represent the "uncertain" parameters as random variables, with the degree of their variability reflecting our level of uncertainty in their estimates. Thus, the value of the uncertain parameter is not held constant during a simulation, and varied from one simulation to another, as in the example above. Rather, the parameter value is drawn randomly from an appropriate statistical distribution or frequency distribution at appropriate times during a simulation, often at each time step during the simulation. As we mentioned in Section 6.5.1, neither option is inherently better than the other; our choice depends, as always, on our specific objectives. In our island population model, changing the value of S during a simulation does not seem to make much sense, since S represents the density-dependent relationship between birth rate and population size. One would imagine that this type of density-dependent relationship would not change from one year to the next in a random manner, at least not over as wide a range of values as we examined in our example.

7.3 **Analyze and interpret simulation results**

As was the case with the development of the experimental design for simulations, the same principles that apply to analysis of real-world experimental data apply to analysis of simulated data. Before we compare our model projections under the different simulated treatments to each other, we first compare them to our a priori expectations, which we described during the final phase of conceptual model formulation. This comparison represents one last evaluation of the model. Just as we may lose confidence in real-world experimental data if we cannot explain aberrant values, we may lose confidence in simulated data if we cannot reconcile differences between these data and our expectations. However, since we are projecting system dynamics under conditions that, presumably, have never been observed in the real world, we

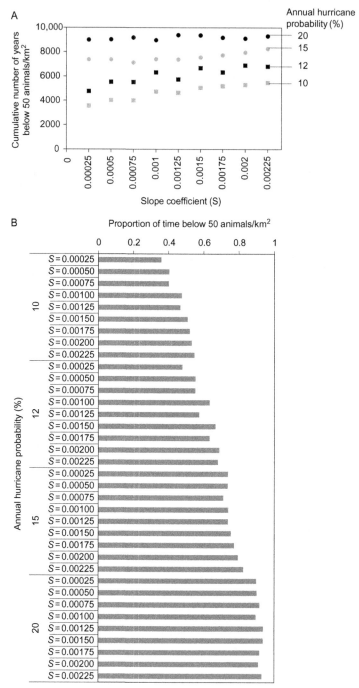

FIG. 7.2

Summary of model application results for the oceanic island population model. (A) The cumulative number of years (during 100 replicate stochastic simulations, each 100 years in length) in which the population size was less than 50 animals per km^2 under the indicated annual hurricane probabilities and values of the slope coefficient (S). (S affects the functional relationship between births and population size: $births = (1 - S \times Pop) \times Pop$). (B) The proportion of time during which the population size was less than 50 animals per km^2 under the indicated annual hurricane probabilities and values of the slope coefficient (S).

should be reluctant to discard model projections as useless at this point. The manner in which we interpret differences between how we imagine (hypothesize) the real-world functions and actual observations of the real world remains problematic, and continues to provide excellent fodder for philosophical debate. Our point is simply that we analyze and interpret simulated data exactly as we would experimental data from the field or laboratory (see Chapter 1, Fig. 1.2).

Assuming we decide to proceed with our analysis, we decide how we will compare results from our various simulations (experimental treatments) the same way we would decide how to compare data from the real system obtained from different experimental treatments. In both cases, our decision is based on our project objectives, the details of our experimental design, and on the type of data we have generated. Again, as we discussed within the context of model evaluation (Chapter 6, Section 6.4), the same restrictions that apply to the use of statistical tests on real-system data (e.g., see Ott and Longnecker, 2001) also apply to model projections (simulated data).

If the model is deterministic, the analysis of system behavior in response to management policies or environmental situations is nonstatistical. Comparison of single-value model predictions consists of ranking the alternative predictions and assessing the practical significance of differences between them. Comparison of times-series predictions consists of assessing the practical significance of differences with regard to the timing of maximum and minimum values and the amplitude and periodicity of fluctuations within the time series. Interpretation of results is solely in terms of the magnitude of practically significant differences established within the context of the particular problem.

If the model is stochastic, analysis of system behavior in response to alternative management policies or environmental situations can be statistical. The statistical comparison of single-value model predictions consists of an analysis of variance testing the hypothesis that the mean of model predictions generated under each alternative scenario is the same. If the analysis of variance indicates statistically significant differences among the alternative scenarios, a multiple comparison (Snedecor and Cochran, 1967, p. 271), such as a Duncan's multiple range test (Ott, 1984, p. 376), can be used to identify the statistically significant differences among specific alternative scenarios. This assumes that more than two alternative scenarios are compared. If only two alternative scenarios are compared, a two-sample t-test can be used, as described in Chapter 6 (Section 6.4.1). The results are interpreted both in terms of the magnitude and the statistical significance of differences in model predictions among alternative scenarios.

The statistical comparison of times-series model predictions is more complicated than is the case for single-value predictions. As discussed in Chapter 6 (Section 6.4.2), exploration of problems associated with the analysis of time-series and the presentation of possible ways to overcome them in specific situations are beyond the scope of this book. The procedure suggested as useful during model evaluation, that is, graphically representing the variance around each mean value in each time series that we wish to compare, is equally useful during model use. But, again, such representation does not imply a statistical test.

As a numerical example of the nonstatistical analysis and interpretation of simulation results, we will return to the results of the model application simulations we conducted with our island population model in the previous section (Fig. 7.2). First, of interest is the fact that the cumulative number of years during which the population size is less than 50 animals per km^2 increases with increasing hurricane probabilities, regardless of the value of S (Fig. 7.2A). That is, our assessment that an increased frequency of hurricanes reduces the ability of the population to sustain itself is robust with regard to our uncertainty in the value of S. This may seem a ridiculously obvious conclusion, and, indeed, in this simple example, it could not be otherwise. However, in more complex models, the model output in which we are most interested may respond in more complex ways to changes in the value of a given parameter. In such cases, it may be difficult to rank the relative effects of different environmental impacts from low to high, or different management schemes from best to worst, because the rankings change depending on the specific parameter value used. Second, also of interest is the relative difference in the ability of the population to sustain itself under any given assumption about hurricane frequency. For example, assuming an annual hurricane frequency of 10%, the population might spend anywhere from about 36% to about 54% of its time below 50 animals per km^2, depending on the value of S used (Fig. 7.2B). On the other hand, assuming an annual hurricane frequency of 20%, the population might spend from about 90% to about 94% of its time below 50 animals per km^2, depending on the value of S used (Fig. 7.2B). Thus, our uncertainty concerning the ability of the population to sustain itself decreases with increasing hurricane frequency. Once again, although this is a ridiculously obvious conclusion for this simple model, in more complex models, our uncertainty in a given model output may respond in more complex ways to changes in the value of a given parameter over the range of environmental conditions or management scenarios in which we are interested.

If we wanted to determine if the differences in the cumulative number of years during which the population size is less than 50 animals per km^2 are statistically significantly different among the different hurricane frequencies, we might conduct a standard one-way analysis of variance (ANOVA) (see, e.g., Ott and Longnecker, 2001). Likewise, if we wanted to determine if the differences in the cumulative number of years during which the population size is less than 50 animals per km^2 are statistically significantly different among the different slope coefficients (among the different S values), we might conduct another standard one-way ANOVA. We have summarized the results of two such ANOVAs in Table 7.1. The ANOVAs indicate that there are highly statistically significant differences ($P < 0.001$) in the cumulative number of years during which the population size is less than 50 animals per km^2 both among the different hurricane frequencies (Table 7.1A) and among the different S values (Table 7.1B). Of course, there are other statistical analyses that we might have conducted depending on our specific interests. For example, we might have conducted a two-way ANOVA including both hurricane frequencies and S values as sources of variation in the cumulative number of years during which the population size is less than 50 animals per km^2. And, we should note that, arguably, for this simple example, these statistical analyses may not have provided us with any additional information of any importance. That is, the differences of interest

Table 7.1 Summary of the analyses of variance (ANOVAs) conducted on results of the model application simulations of the island population model. (A) ANOVA of the differences in the cumulative number of years during which the population size is less than 50 animals per km² among the different hurricane frequencies (annual probabilities). (B) ANOVA of the differences in the cumulative number of years during which the population size is less than 50 animals per km² among the different slope coefficients (among the different S values).

(A)

Hurricane probability (%)	N	Sum	Average	Variance
10	900	42,022	46.69111	512.0513
12	900	54,587	60.65222	547.3305
15	900	68,083	75.64778	465.2607
20	900	82,390	91.54444	151.79

Source of variation	SS	df	MS	F	P-value	F crit
Between groups	1,007,353	3	335784.2	801.1875	0	2.607379
Within groups	1,507,113	3596	419.1081			
Total	2,514,465	3599				

(B)

S-value	N	Sum	Average	Variance
0.00025	400	24,691	61.7275	990.8203
0.0005	400	25,930	64.825	858.641
0.00075	400	25,814	64.535	801.9035
0.001	400	27,410	68.525	668.636
0.00125	400	27,133	67.8325	688.1598
0.0015	400	28,640	71.6	579.6942
0.00175	400	28,414	71.035	555.1667
0.002	400	29,237	73.0925	521.2872
0.00225	400	29,813	74.5325	488.3398

Source of variation	SS	df	MS	F	P-value	F crit
Between groups	59558.72	8	7444.839	10.8902	2.8E-15	1.940981
Within groups	2,454,907	3591	683.6276			
Total	2,514,465	3599				

probably were clear enough before conducting the ANOVAs. As we mentioned in Chapter 6 (Section 6.4.4), there is a tendency to think that quantitative methods, particularly those involving statistical tests, are more rigorous and definitive than qualitative methods. However, quantitative methods are not inherently better than qualitative methods, and both should be viewed within a framework that emphasizes the usefulness of the information provided to meet the objectives of the modeling project.

7.3.1 Examine additional types of management policies or environmental situations

At this point, we should have at least tentative answers to initial questions specified during conceptual-model formulation. However, the initial simulations often raise as many questions as they answer. We may learn something new about the behavior of the system-of-interest through simulation that suggests a new line of questions that are relevant to our interests or that allow us to refine our original questions to obtain more precise answers. This is exactly the iterative procedure of analysis and redesign on which all scientific experimentation is based. The new set of questions that we wish to address may necessitate fundamental changes in our model or may even suggest that we must redefine our system-of-interest. In such cases, we return to the first phase of systems analysis, conceptual-model formulation, and repeat the entire model-building procedure. But often our new questions can be addressed using essentially the same model, slightly modified to represent a new variant of the management policies or environmental situations that we represented in our original model-application simulations.

Often, this step in model application consists of two parts: (1) an exploratory examination of system behavior under relatively many variants of a few original management policies or environmental situations and (2) a detailed description and/or formal comparison of system behavior under selected new variants. During exploratory simulations, we are interested primarily in identifying trends in system behavior and not in the detailed description or the formal comparison of system behavior under all of the management or environmental variants simulated. If we have a stochastic model, it often is more efficient to conduct these exploratory simulations using a deterministic version of the model in which all random variables are assigned their mean values. The trends in system behavior are clearer without the inherent variability of the stochastic model and fewer simulations are required. After examining the results of the exploratory simulations, we may identify certain new management policies or environmental situations that we wish to describe in detail and/or compare formally. The formal comparison requires the development of a new experimental design for simulations. For deterministic models, we may have run most or all simulations required by the new experimental design as part of the exploratory analysis. But for stochastic models, we must recalculate the sample size, or the number of replicate stochastic simulations required by the new experimental design, run the replicate stochastic simulations, and conduct the appropriate statistical analyses, as described in the previous section.

As a numerical example of examining additional types of management policies or environmental situations, we will return once again to the results of the model application simulations we conducted with our island population model in the previous section. We had concluded that the cumulative number of years during which the population size is less than 50 animals per km^2 increases with increasing hurricane probabilities, regardless of the value of S (Fig. 7.2A), and that our uncertainty concerning the ability of the population to sustain itself decreases with increasing hurricane frequency (Fig. 7.2B). In addition to these conclusions, we also might have noticed that the fate of the simulated population depends primarily on the size of the population when the first hurricane occurs, and on whether or not hurricanes occur in consecutive years.

Based on these observations, we might decide to run some additional simulations exploring various combinations of population size when the first hurricane occurs and on whether or not hurricanes occur in consecutive years. We probably would use a deterministic version of our model in which we prescribe a priori exactly when hurricanes would occur. For example, if the first hurricane occurs in year 5 of simulated time, when the population size is 47.35 animals per km^2, and no other hurricanes occur until after the population has reached carrying capacity (400 animals per km^2, assuming that $S = 0.00125$), the population is reduced to 4.74 animals per km^2, after which it takes the population approximately 20 years to reach carrying capacity (Fig. 7.3). If the first hurricane occurs in year 5 and is followed by a second hurricane

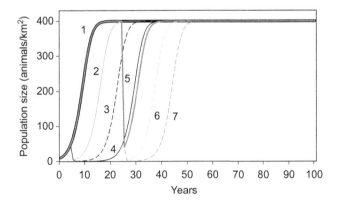

FIG. 7.3

Summary of results of simulating additional types of hurricane frequencies with the oceanic island population model. The seven lines on the graph depict population dynamics during simulations in which (1) no hurricanes occurred (series 1 line), (2) the first and only hurricane occurred in year 5 of simulated time (series 2 line), (3) the first hurricane occurred in year 5 and was followed by a second hurricane in year 6 (series 3 line), (4) the first hurricane occurred in year 5 and was followed by hurricanes in years 6 and 7 (series 4 line), (5) the first hurricane did not occur until year 25 (series 5 line), (6) the first hurricane occurred in year 25 and was followed by a second hurricane in year 26 (series 6 line), and (7) the first hurricane occurred in year 25 and was followed by hurricanes in years 26 and 27.

in year 6, the population is reduced to 0.47 animals per km^2, after which it takes the population approximately 26 years to reach carrying capacity. If the first hurricane occurs in year 5 and is followed by hurricanes in years 6 and 7, the population is reduced to 0.047 animals per km^2, after which it takes the population approximately 33 years to reach carrying capacity.

On the other hand, if the first hurricane does not occur until year 25 of simulated time, when the population already is at carrying capacity, and no other hurricanes occur until after the population has returned to carrying capacity, the population is reduced to 40.0 animals per km^2, after which it takes the population approximately 16 years to return to carrying capacity (Fig. 7.3). If the first hurricane occurs in year 25 and is followed by a second hurricane in year 26, the population is reduced to 4.0 animals per km^2, after which it takes the population approximately 22 years to return to carrying capacity. If the first hurricane occurs in year 25 and is followed by hurricanes in years 26 and 27, the population is reduced to 0.4 animals per km^2, after which it takes the population approximately 28 years to return to carrying capacity.

Based on the results of these exploratory simulations, it might occur to us, for example, that it would be interesting to see what effect introducing a hurricane refuge on the island might have. That is, a refuge that would provide shelter to a certain number of animals when a hurricane occurred, thus preventing population size from being reduced so drastically. This new interest might provide the rationale for the development of a new experimental design for simulations which would allow the formal comparison of the effects of refuges of various sizes on the ability of the population to sustain itself. One could imagine repeating our original 36-scenario experimental design (see Section 7.2) several times, each time with the addition of a refuge of a different size.

7.4 Communicate simulation results

The final step in model use involves communication of simulation results. Within a research setting, this usually means effective communication of the modeling methodology as well as the model results to research scientists via publication in an appropriate scientific journal. Within a management framework, this usually implies effective communication of model results to natural resource managers and other groups interested in the modeled system (i.e., to stakeholders), as well as to policy makers whose decisions ultimately impact natural resources. As with the design, analysis, and interpretation of simulated experiments, the communication of simulation results is no different than communication of results of real-world experiments. We must describe the problem we are addressing, our specific objectives, the information base we are drawing upon, the technical method used to analyze the information (i.e., we must formally describe the model), the results of evaluation procedures and (simulated) experiments, and our conclusions. However, the relative emphasis placed on these various aspects of the information to be communicated, as well as the style of communication, depends on the target audience. Although the details regarding the most appropriate content and the most appropriate style will differ from one situation to another, we might identify two general types of target audiences: (1) research scientists, and (2) natural resource managers and stakeholders. (Students and the general public form a third important group. However, a discussion of the use of simulation models in educational and public outreach settings, although these uses are important and widespread, is beyond the scope of this book.)

7.4.1 Communication within a research setting

Effective communication of both the modeling methodology and model results to research scientists has been facilitated greatly by recent technological advances that allow electronic archiving of model code as well as the data used to support model development. However, these technological advances do not lessen the need for modelers to describe their models well. By "describe well" we mean not only to describe precisely and unambiguously the equations and formal logical statements that comprise the model, but also the assumptions upon which the model is based and the larger context within which the model has been evaluated and can be used appropriately. By "context" we mean the domain of applicability of the model as described by Rykiel (1996). That is, the context of a model is the domain over which it is intended to operate, which embodies all of the assumptions under which it operates, and refers especially to those assumptions that remain implicit. We will consider further the importance of this concept of model context when we reflect on "how good (valid) are models?" in Chapter 10 (Section 10.7.1), and also when we reflect on the potential dangers of the "integration of existing models" in Chapter 13 (Section 13.4). For now, suffice it to say that the context of a model should be described clearly enough to prevent, or at least minimize, the possibility that it might be applied to problems outside the domain over which it is intended to operate.

Regarding a format for effective communication of both modeling methodology and model results to research scientists via publication in a scientific journal, we find useful the general format suggested by Grant and Swannack (2008). These authors observed that the description of simulation modeling projects using the typical "methods" and "results" format most commonly found in scientific ecological journals often seems awkward. A simulation model description might be viewed as either a part of the project methodology or as a part of the project results. Thus, these authors suggested replacing the "methods" and "results" section headings with sections titled "overview of the model," "model description," "model evaluation," and "simulation of [...]," with the latter referring to scenarios that were simulated to meet model objectives. These sections would follow the typical "introduction" and precede the typical "discussion" sections. Of course, some journals have rigid formats that require "methods" and "results" section headings. In such journals, "overview of the model" and "model description" might be subheadings under "methods," and "model evaluation" and "simulation of [...]" might be subheadings under "results." However, in some cases, the "model evaluation" material might be placed more appropriately as a subheading under "methods."

Below, as an example of use of the general format suggested by Grant and Swannack (2008), we present a "bare bones" description our stochastic island population model.

Title: Effect of hurricane frequency on sustainability of an oceanic island animal population

Introduction: There is concern about the future of an animal population on an isolated oceanic island that is subjected to hurricanes. In particular, in view of a projected increase in the frequency of hurricanes over the next few decades, there is concern over the number of years it will take for the population to recover to its prehurricane size. The animal population that inhabits the island has been reduced drastically in size following a recent large hurricane. The most recent study estimated that there only are about 10 animals per km^2 remaining in the population. In this paper, we describe a model that simulates the effect of variable hurricane frequency on the ability of the population to recover to its prehurricane size.

Overview of the model: The model represents the population dynamics of an oceanic island animal population in response to the frequency of hurricanes (Fig. 6.4A). The number of animals in the population changes annually as a result of births and deaths. There is no immigration or emigration to or from the island. Birth rate is density-dependent, and both birth rate and death rate are affected by hurricanes.

Model description: We developed the model as a compartment model based on difference equations ($\Delta t = 1$ year) and programmed in (identify software). Population density (*Pop*, animals km^{-2}) changes as a result of the net difference between births (*b*, animals km^{-2} year^{-1}) and deaths (*d*, animals km^{-2} year^{-1}). Both *per capita* birth rate (*br*, animals animal^{-1} year^{-1}) and *per capita* death rate (*dr*, animals animal^{-1} year^{-1}) are affected by hurricanes (*h*, unit-less index). *Per capita* birth rate also is affected by population density.

$$Pop_{t+1} = Pop_t + (b_t - d_t)\,\Delta t \tag{7.1}$$

$$b_t = br_t \times Pop_t \tag{7.2}$$

$$d_t = dr_t \times Pop_t \tag{7.3}$$

$$br_t = (1 - \gamma \times Pop_t) \times Pop_t, \text{ if } h_t = 0 \tag{7.4}$$

$$br_t = 0, \text{ if } h_t = 1 \tag{7.5}$$

$$dr_t = \delta, \text{ if } h_t = 0 \tag{7.6}$$

$$dr_t = \mu, \text{ if } h_t = 1 \tag{7.7}$$

where $h_t = 0$ indicates that no hurricanes occurred in year t, $h_t = 1$ indicates that a hurricane occurred in year t, and γ, δ, and μ are constants. Available data suggest that during nonhurricane years *dr* has been 50%, thus $\delta = 0.5$, whereas during hurricane years *dr* has been 90%, thus $\mu = 0.9$. Analysis of historical records suggests that *br* decreases linearly from a maximum of 1 to a minimum of 0 as population size increases from essentially zero to 800 per km^2, thus $\gamma = 0.00125$. Historically, hurricanes have swept over the island about once every 10 years. Thus h_t is drawn randomly each year during a simulation from a Bernoulli distribution with $p = 0.1$. Finally, for the baseline simulations, $Pop_0 = 10$, based on the most recent posthurricane study.

Model evaluation: To evaluate the usefulness of the model in meeting our objectives, we assessed model structure as well as model behavior under the baseline conditions. That is, during nonhurricane years. Model structure seems reasonable in that, given the objective of the model and the data available, we are interested in increases in population size resulting from the net difference between annual births and deaths. Model behavior also seems reasonable. Based on population growth theory, we know that populations growing in a limited environment should exhibit sigmoidal (S-shaped) growth. That is, population size should increase at an increasing rate until the population reaches one-half of the carrying capacity of the environment, and then should increase at a decreasing rate until it reaches carrying capacity. We assume that we are dealing with a closed population, that is, there is no immigration nor emigration, which seems reasonable given the model simulates a population on an isolated oceanic island.

(For the sake of brevity, we will not describe the sensitivity analysis that we conducted in Chapter 6 (Section 6.5), and we will conduct our model application simulations (below) using only the baseline version of the model.)

Simulation of effects of hurricane frequency on population sustainability: We will assess the effect of an increased frequency of hurricanes on the ability of the island population to sustain itself by simulating four scenarios in which the annual probabilities of occurrence of hurricanes will be 10%, 12%, 15%, and 20%. We will run 100 replicate stochastic simulations of each of these scenarios, and, during each simulation, we will monitor the cumulative number of years during which the population size is less than 50 animals per km^2.

Simulation results suggest that the ability of the population to sustain itself, as indicated by the proportion of simulated time during which the population size is less than 50 animals per km^2, decreases with increasing hurricane frequency (Fig. 7.2B).

Discussion: The current concern about the future of the animal population on the isolated oceanic island, in view of the projected increase in the frequency of hurricanes over the next few decades, seems well justified. The proportion of time that the population size would be maintained below the critical threshold of 50 animals per km^2, given even a relatively moderate increase in the annual probability of hurricanes from 10% to 15%, would increase from less than 50% to more than 70%.

7.4.2 **Communication within a management framework**

Effective communication of model results to natural resource managers and stakeholders, that is, to potential end users of the model, is facilitated greatly by their early involvement in the modeling project. Ideally, natural resource policy makers also would be included in this group, although most often this is not feasible. Model development usually requires many subjective decisions that, when viewed en masse by users for the first time only after completion of the model, can be quite difficult to explain. User confidence in the model, as well as the overall quality of the model, almost always is higher as the result of early and continued dialogue with the potential end users. Communication also is easier if emphasis is placed on interpreting general trends in model behavior in ecological terms. The general trends usually are of more relevance within a management context than are the specific numerical values of model projections. The end users, as well as the modelers, also have more confidence in the ability of the model to predict general trends and almost always feel less comfortable with any given single numerical projection. However, for most modelers, there is a tendency even with relatively simple models to become preoccupied with presenting detailed results at the expense of a clear overview. This tendency is ironic considering that an underlying rationale for use of a modeling approach is to attain a holistic view of our problem.

Cartwright et al. (2016) have suggested a set of critical elements that facilitate effective communication of complex ecological models to nonscientist end users (Fig. 7.4). Modelers initially define the message to be communicated. Essential components of this message should include model predictions and the associated uncertainties. Complexity of the message should depend on the context of the message,

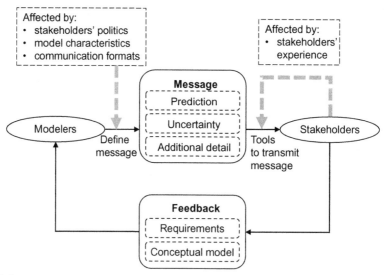

FIG. 7.4

Illustration of a set of critical elements that facilitate effective communication of complex ecological models to nonscientist end users. See text for details.

Modified from Cartwright, S.J., Bowgen, K.M., et al., 2016. Communicating complex ecological models to non-scientist end users. Ecol. Model. 338, 51–59.

the complexity of the model, and the method of communication. Types of communication tools that could be employed might include visualizations and real-time use of model graphical interfaces. Both before and after the message has been conveyed, end users should indicate to the researchers what their information requirements are, which should help inform overall model design. Cartwright et al. (2016) also have provided an excellent overview of the challenges associated with effective communication of complex ecological models to nonscientist end users, and have suggested a general approach to facilitate effective communication with this group. We have drawn heavily upon their work in the following discussion.

Challenges associated with the effective communication

Cartwright et al. (2016) associated challenges of communicating the content of complex ecological models to end users with: (1) political context of the problem being modeled, (2) experience of end users with modeling technology and with use of outputs from models in decision-making, (3) the particular characteristics of the model, (4) choice of the most appropriate manner to represent the uncertainty associated with model outputs, and (5) the particular form of communication best suited to the end users.

Political context of the problem being modeled is of particular importance when critical real-world decisions are at stake. In such situations, the modeling process itself can become politicized, and communication of model outputs pertaining to highly politicized problems requires careful consideration to maximize stakeholder trust and minimize misunderstanding or misinterpretation of model results. In highly polarized debates, end users often come from opposing sides, and entrenched opinions can lead to responses whereby end users either believe model outputs uncritically (with "blind faith") or completely reject all simulation results. Uncritical acceptance of model outputs poses the risk that end users will believe the model implicitly, without understanding the uncertainties and limitations associated with model results. Ultimately, this might lead to poor decisions being made if model outputs are trusted beyond the domain of validity of the model. On the other hand, if end users refuse to accept simulation results that contradict their point of view, this raises the question of the extent to which it is the responsibility of modelers to ensure that model results are incorporated into the decision-making process appropriately.

Experience of end users with modeling technology and with use of outputs from models in decision-making can affect the communication process. End users may have preexisting ideas about computer simulations as opposed to "real-world" empirical data, and thus will differ in their practical requirements for receiving information about the model and model outputs. For example, some of end users may prefer simple mathematical models with which they are familiar. Often these are models that are widely employed because nonscientists can understand them, but which are completely inappropriate for prediction. These end users may be inherently skeptical of predictions arising from complex models, particularly if outputs of these models contradict current opinion or practice, or appear superficially counterintuitive. They may be unwilling to learn enough about a complex model to understand the true degree of uncertainty associated with model outputs. Alternatively, some end users may assume that a complex model is a complete representation of the real world, rather than a tool to simulate key processes and test hypotheses that cannot be tested easily in the laboratory or the field. In such cases, the risk is that model outputs may be trusted uncritically and interpreted incorrectly. The challenge for modelers is to achieve the middle ground, that is, to achieve informed acceptance of model outputs. Finally, some end users may be exposed routinely to complex models and will be accustomed to technical descriptions and will require a succinct approach in a familiar format which will enable information to be delivered, understood, and acted upon rapidly.

The particular characteristics of the model will affect the communication approach taken. For example, single-species models that address simple systems may be easier to communicate than multi-species, multi-trophic level models that address multidimensional problems. Obviously, modelers should not develop a model that is too complex or too simple for the problem at hand. However, modelers can fall easily into the trap of putting "everything" into the model, or of passively accepting a gradual ratcheting-up of model complexity as a result of peer pressure to have a more comprehensive model, which superficially may appear to be more credible. Modelers also can succumb to the temptation to push the capabilities of modeling software as far as possible, irrespective of the need to do so, with the resulting risk of producing an overly complex model which is unintuitive, difficult to understand, and viewed as a "black box." Likely, such a model will contain parameters that are difficult to justify, buried assumptions that are difficult to detect, and hidden errors that, in fact, have not been detected. On the other hand, overly simple models, which may help to frame the problem at hand conceptually, often contain simplistic variables of doubtful relevance and omit important system processes. Once again, the challenge for modelers is to achieve the middle ground. In this instance, the middle ground represents the balance between model sophistication and model simplicity that results in development of a useful decision-support tool.

Choice of appropriate manners to represent the uncertainty associated with models is crucial with regard to increasing end user confidence in, and decreasing likelihood of misuse of, model results. Uncertainty in predictions of complex models originates from imprecision in parameter estimates, inherent variability in the system being modeled, uncertainty associated with model structure, and uncertainty regarding future scenarios that the model will be used to simulate. Any uncertainty associated with model output can reduce end user confidence. However, presenting uncertainty to end users in ways that are natural to modelers, such as confidence limits of a continuous probability distribution, could lead to false end user confidence that the response of the system is captured fully within the limits of this range. A probability distribution can be difficult to translate into the discrete options that a decision-maker requires.

Choice of the particular form of communication that is best suited to end users also is crucial with regard to the perceived utility of model results. Communication format and communication timeframe often are determined by the end user audience, irrespective of the format and timeframe that modelers would prefer. When direct communication with end users is limited to short, infrequent time slots, important explanations of model outputs and their implications risk being summarized to the point of irrelevance, or over-looked completely. Using technical language, or failing to explain the implication of model results in lay terms, also is a barrier to effective communication, but often occurs if modelers fail to gauge correctly the end users' level of technical familiarity with the subject matter of the model. For example, when conveying uncertainty, end users may understand "uncertainty" as reflecting a lack of knowledge on the part of the modelers rather than as a genuine property of model predictions. If the end users interpret prediction uncertainty as a lack of knowledge on the part of the modelers, then any technical language used to convey uncertainty may be perceived as an attempt to conceal the ignorance of the modelers, thereby undermining credibility of the model and reducing end user trust. Nonscientists also may struggle to understand how predictions delivered as percentage likelihoods actually could be tested against empirical data, thus making it important to explain to end users precisely how likelihood predictions were validated.

A general approach to facilitate effective communication
Cartwright et al. (2016) suggested a general four-stage process for effectively communicating the content and outputs of complex ecological models to end users, which included: (1) involving end users during model development, (2) preparing the message to be communicated with regard to what end users most care about, (3) choosing the formats within which communication will take place, and (4) evaluating effectiveness of the communication process.

Involving end users during model development is important in order to identify their requirements and expectations. Ideally, end users should be involved throughout the modeling process. They should help identify the practical aims of the model and help to formulate the conceptual model. Any ambiguity at this stage of modeling process risks the development of inappropriate models and the misinterpretation of model outputs. Achieving consensus on the conceptual model and clarity on the practical aims of the model increases end users' familiarity, trust, and investment in the modeling process. End users should help determine degree of model complexity and entities included in the model, as well as quantities included in the model output. They also can provide key insights into parameter values and model processes that are unknown to modelers, and which are unavailable through the scientific literature. End users should see the results of testing each new submodel, and they should be shown the model's foundations in established theory and the scientific literature, and its resemblance to other trusted models. By incorporating a user-friendly model interface, modelers can provide end users with the ability to adjust selected parameter values themselves and see how their adjustments affect model results without needing to fully understand the inner workings of the model.

Preparation of the message to be communicated to end users should be focused on what end users most care about, and also on questions that modelers anticipate end users are most likely to have about the model. Communicating effectively requires an understanding of end users and the tailoring of the communication format to meet their needs. In order to do this, modelers must consider the backgrounds of end users. For example, do they have an agenda? What language will they understand? What narratives will resonate with them? What will affect their decisions? How can you prepare models of realistic scenarios that will be relevant to their decisions, and how can you convey results of those modeled scenarios in a way that end users will understand? For example, if model predictions are expressed as a probability distribution but the decisions of end users must be based on evaluation of alternative options, modelers could consider placing these alternative options onto a continuous probability scale so that it is easy for end users to see the probabilities associated with each option. If model predictions lead to solutions involving beyond-business-as-usual scenarios, then part of the communication process requires introduction of end users to these unexpected or novel solutions. In such cases, multiple informal discussions to familiarize end users with the new ideas will be important. Finally, in order to anticipate criticisms and challenges of end users, it is essential for modelers to know the model thoroughly, including how model components relate to each other, assumptions and generalizations involved in the model and their justifications, the data used to validate the model, the specific contexts in which the model is valid, and how the uncertainty associated with model outputs can be presented honestly and comprehensibly. Modelers must thoroughly document and justify each stage in the modeling process and be prepared to demonstrate that they are aware of, and have incorporated, the relevant literature and expert opinion into the model.

Choosing the formats within which communication will take place is critical to effective communication. Multiple communication formats help fulfill different aims. For example, to communicate an overview of the purpose and the outputs of a model, a meeting or workshop attended by relevant end users, followed by a question and answer session, may be appropriate. This could include dynamic visualizations of model output, which could be prerecorded. Using a graphical interface as the model front end is invaluable to visualize model behavior as it runs, and can show the effects of altering parameter values and input values for management scenarios in real time. Experienced end users then can assess whether the model responds according to their expectations. This can help resource managers understand the system they manage, and also can affect whether or not well-informed end users believe the model results. Initial meetings and workshops can be followed by informal one-on-one discussions of model details. The model also can be made available for Internet download with a guided tour and manual that end users can explore at their leisure.

Evaluating effectiveness of the communication process has received relatively little formal attention, although the degree to which the communication process is effective ultimately governs the relevance and societal impact of the model. At present, there is little understanding of what constitutes effective communication with end users, let alone how to measure the effectiveness. Three possible criteria for judging whether communication of a complex model and it outputs to end users is effective might include: (1) Do end users subsequently have enhanced knowledge of the model and its relevance? (2) Does the model affect subsequent management decisions? (3) Does the behavior of the system being modeled consequently improve? During early stages of the modeling process, it may only be appropriate to measure the effectiveness of communication based on the first criterion. One immediate, qualitative measure of effectiveness is the intensity of discussion following direct communication, such as a presentation at a meeting. Lengthy discussion with many questions might suggest that the audience was engaged and cares about the information presented. However, silence following a presentation is not necessarily bad, particularly when end users have been exposed to novel scenarios or modeling methods. The measurement of the effectiveness of communication based on the second and third criteria may not be appropriate for several to many years.

Some practical considerations

PART

3

Potential pitfalls during the modeling process

Chapter outline

Developments in Environmental Modeling, Volume 31, ISSN 0167-8892, https://doi.org/10.1016/B978-0-444-64163-2.00008-6

In this chapter, we point out some of the pitfalls that are commonly encountered during the model development process. Although many, if not most, of these pitfalls may sound almost trivially silly when put into words, we find ourselves continually pointing them out to beginning modelers. And, the truth be known, we still find ourselves at the bottom of one of these pits from time to time, wondering why we fell. In fact, the majority of the pitfalls that we will note below might well qualify as "historic landmarks" on the modeling landscape, since most were identified over four decades ago by Jay W. Forrester in his classic work on modeling industrial dynamics (Forrester, 1961). Thus, following humbly in the footsteps of Professor Forrester, as a complement to our practical "how to" modeling guide, which we will present in the next chapter (Chapter 9: A candid view of the modeling process), we offer the following "how not to" modeling guide. We have organized these pitfalls according to the phase during the modeling process in which we most commonly encounter them (Table 8.1).

8.1 Pitfalls during conceptual model formulation
8.1.1 State the model objectives

P1. Inadequate definition of model purpose: We should design our model to address specific questions. If our questions are stated vaguely, our decisions regarding what to include in the model will be inconsistent and the model structure will be flawed. Model objectives beginning with "To understand ..." almost always lead to trouble.

P2. Implicit criteria for model evaluation: We should specify precisely the criteria our model must meet to be considered useful. If our evaluation criteria remain implicit, and hence vague, our statements regarding the usefulness of the model will be equally vague and will lack credibility.

P3. No description of model context: We should describe the context within which we intend the model to operate. If we do not list the restrictive assumptions regarding the real-world conditions under which our model is useful, our model may be applied inappropriately and provide erroneous projections. A frequent complaint of modelers is that their model appears flawed because it has been applied to problems that it was not designed to address. The failure to describe model context invites such misuse.

8.1.2 Bound the system-of-interest

P4. Casual choice of the scale for the system-of-interest: We should give careful thought to the manner in which we bound the system-of-interest in time and space. If we bound the system too narrowly, we risk losing the possibility to simulate those situations that would be most helpful in solving our problem. If we bound the system too broadly, we risk focusing too much attention on irrelevant components and processes, and failing to recognize the interactions most important for solving our problem. For example, we probably could not adequately represent the effects of

Table 8.1 Some common pitfalls encountered during the four phases of model development.

	Theory																		
	I: Conceptual model formulation						**II: Quantitative model specification**					**III: Model evaluation**				**IV: Model application**		
Pitfall	a	b	c	d	e	f	a	b	c	d	e	a	b	c	d	A	b	c
(1) Inadequate definition of model purpose	✓																	
(2) Implicit criteria for model evaluation	✓																	
(3) No description of model purpose	✓																	
(4) Casual choice of scale of the system-of-interest		✓																
(5) Inclusion of too many components		✓																
(6) Careless categorization of system components			✓															
(7) Inclusion of excessive detail				✓														
(8) Inclusion of circular logic				✓														
(9) Lack of precision in conceptual model diagram					✓													
(10) Reluctance to make initial hypotheses about system behavior						✓												
(11) Selection of inappropriate mathematics							✓											
(12) Selection of inappropriate software							✓											
(13) Choice of inappropriate time unit for simulations								✓										
(14) Construction of mathematical descriptions without clear meaning									✓									
(15) Underestimation of the importance of graphical representations									✓									
(16) Use of functional relationships that are not interpretable									✓									

Continued

Table 8.1 Some common pitfalls encountered during the four phases of model development.—cont'd

Pitfall	I: Conceptual model formulation						II: Quantitative model specification					III: Model evaluation				IV: Model application		
	a	b	c	d	e	f	a	b	c	d	e	a	b	c	d	A	b	c
(17) Carelessness in defining dimensional units of model components										✓								
(18) Use of coefficients without meaning to obtain dimensional consistency										✓								
(19) Reluctance to use qualitative information										✓								
(20) Decision to remove a functional relationship due to lack of data										✓								
(21) Reliance on automated model parameterization										✓								
(22) Use of overly-sophisticated equations										✓								
(23) Underestimation of the importance of negative feedbacks and time lags										✓								
(24) Careless definition of baseline conditions											✓							
(25) Reliance on automated solutions to mathematical and programming problems											✓	✓						
(26) Underestimation of the importance of qualitative aspects of model evaluation												✓						
(27) Acceptance of conceptually flawed functional relationships													✓					
(28) Acceptance or rejection of surprising model behavior without explanation																		

(29) Inappropriate interpretation of the initial phase of model behavior

(30) Reliance on automated model evaluation

(31) Tendency to believe all data from the real system are correct

(32) Careless interpretation of statistical tests used in model evaluation

(33) Careless design of sensitivity analysis

(34) Tendency to equate model sensitivity with model inadequacy

(35) Careless experimental design for model application

(36) Tendency to overestimate the range of model applicability

(37) Inappropriate interpretation of statistical tests used in model application

(38) Failure to communicate numerical results directly in ecological terms

(39) Tendency to imply false expectations regarding model projections

Modified from Forrester, J.W., 1961. Industrial Dynamics. Pegasus Communications, Waltham, MA.

global climate change on the land cover in the British Isles by focusing on 1 ha of land over a period of 1 year, or by focusing on shifts in the orbit of the earth around the sun over a period of 1 billion years.

P5. The inclusion of too many components: We should resist the tendency to include in the model every component that comes to mind as we think about our problem. The criteria for including a component should be equally rigorous as the criteria for exclusion. The inclusion of too many components can hinder our recognition and appropriate representation of the most important interactions.

8.1.3 Categorize components within the system-of-interest

P6. Careless categorization of the system components: We should be rigorous in our categorization of the components within the system-of-interest. Each type of component plays a different role in model structure. Components that appear similar in the real system need not necessarily play similar roles in terms of model structure. For example, representing components as state variables when they may be better represented as auxiliary variables; a model addressing certain questions about predator population dynamics might appropriately represent both the predator and prey populations as state variables, whereas a model addressing different questions about predator population dynamics might more appropriately represent the predator population as a state variable and the prey population as a driving variable (assuming changes in the prey population were not affected by any other components in the modeled system). Careless categorization of system components, along with a lack of precision in the conceptual model diagram (*P9*), are two of the most common pitfalls for beginning modelers.

8.1.4 Identify relationships among components of interest

P7. Inclusion of excessive detail: We should resist the tendency to represent the components and processes in our model with great detail, just because we have the knowledge to do so. The inclusion of too many details can obscure important cause-effect relationships that operate at the more aggregated levels of organization, that is, we lose sight of the forest because we cannot see past the trees.

P8. Inclusion of circular logic: We should scrutinize the conceptual model continually in search of circular logic. As the connectedness of the model increases, we risk adding circular connections (via information transfers) among the auxiliary variables and material transfers. That is, for example, we represent A as a function of B, B as a function of C, and C as a function of A.

8.1.5 Represent conceptual model

P9. Lack of precision in conceptual model diagram: We should familiarize ourselves with the precise definitions of the different types of model components (constant, driving variable, auxiliary variable, state variable, material transfer, information transfer) and use the symbols that represent these precisely defined components to construct our conceptual models. Imprecise description of the conceptual model greatly increases the likelihood of categorizing system components carelessly (*P6*) and of including circular logic in the model structure (*P8*). Imprecise description of the conceptual model also greatly increases the likelihood of subsequent errors during the model quantification phase. Lack of precision in the conceptual model diagram, along with carelessly categorizing system components (*P6*) are two of the most common pitfalls. Beginning modelers should spend ample time becoming familiar the definitions of the various types of model components before they begin conceptualizing their system of interest.

8.1.6 Describe expected patterns of model behavior

P10. Reluctance to make initial hypotheses about system behavior: We should formulate initial written hypotheses about all aspects of the behavior of the system-of-interest. We should avoid the tendency to view these a priori hypotheses as the answers that the model will give us. Rather, we should view them just as we view the hypotheses that we make before any laboratory or field experiment. Our initial written hypotheses, which often are accompanied by graphical representations, serve as important heuristic reference points during the subsequent phases of model development and model evaluation.

8.2 Pitfalls during quantitative model specification
8.2.1 Select the general quantitative structure for the model

P11. Selection of inappropriate mathematics: We should choose the mathematics that we use to represent the model carefully (or consult with an experienced simulation modeler). Not all mathematical formats lend themselves equally well to the representation of a given problem. We risk the solution of the model becoming a mathematical exercise in which some steps are devoid of ecological interpretation, thus reducing the credibility of model results.

P12. Selection of inappropriate software: We should choose the software that we use to program the model carefully (or consult with an experienced simulation modeler). This is closely related to the choice of the mathematics that we use to represent the model (*P11*). Not all computer languages and simulation programs lend themselves equally well to the representation of a given problem. We risk the solution of the model becoming more influenced by the constraints of the computer language or simulation program that we use than by the ecological relationships that we are trying to represent, thus reducing the credibility of model results.

8.2.2 Choose the basic time unit for the simulations

P13. Choice of an inappropriate time unit for the simulations: We should use great care in selecting the basic time unit for our simulations. If our time unit is too long, we risk violating the assumption that all rates of change in the system can be viewed as constant during any given time unit. This can preclude the possibility of appropriately representing negative feedback within the system (see *P23*), which can produce artificial instability in the model. If our time unit is too short, we risk reducing the interpretability, and, hence, the credibility, of the model, and we also will increase the length of time needed to run the required simulations unnecessarily.

8.2.3 Identify the functional forms of the model equations

P14. Construction of mathematical descriptions without clear verbal descriptions: We should describe the functional relationships in the model clearly in words before we attempt to describe them mathematically. If we underestimate the importance of verbal descriptions, we risk being precisely wrong in our mathematical formulations. That is, we should use the precise, but inflexible, language of mathematics to "fine tune" our general, flexible verbal descriptions. Our initial written hypotheses about the system behavior (*P10*) should provide a good point of departure for our verbal descriptions. We have found that a good rule of thumb is to try to explain the model verbally within a couple of minutes. If there are areas of the model that are difficult to explain, they will more than likely be difficult to model mathematically. These are generally the areas that require more research and more thought.

P15. Underestimation of the importance of graphical representations: We should describe the functional relationships in the model graphically whenever possible. Graphical representations often provide a natural intermediate step between verbal representations and mathematical representations. If we underestimate the usefulness of graphical relationships, we risk struggling unnecessarily with the translation of verbal representations into mathematical descriptions.

P16. Use of functional relationships that are not interpretable: We should be able to interpret functional relationships within the subject matter context of our problem, for example, ecologically, physiologically, and so on. Obviously, this applies only to those explanatory models for which such interpretability is an objective. If we cannot interpret some of the functional relationships, due to the connectedness of model components, we risk losing the ability to describe overall behavior of the model in terms of cause and effect.

8.2.4 Estimate parameters of the model equations

P17. Careless definition of the dimensional units of the model components: We should take care in defining all model components precisely, including appropriate units of measure. If we are careless in our definitions, we risk making erroneous calculations due to dimensional incompatibility among model parameters, thus creating nonsensical variables which reduce credibility of the model.

P18. Use of coefficients without meaning to obtain dimensional consistency: We should not create meaningless coefficients as a means of obtaining dimensional consistency among model components. The perceived need to create such coefficients results directly from the careless definition of units of measure of model components (*P17*).

P19. Reluctance to use qualitative information: We should not overlook the value of qualitative information in quantifying relationships within the model. If we try to rely solely on available data to quantify our entire model, we risk a long and probably futile search for numerical information that quite likely does not exist. We almost always know more about the system-of-interest than we can confirm rigorously with available data. We should not hesitate to create arbitrary scales of measurement (often unit-less indexes) to convert qualitative information to quantitative values; quantification does not imply accuracy, only precision. Once we have quantified a relationship, we can then determine much about the required numerical accuracy.

P20. Decision to remove functional relationships from the model due to a lack of data: We should resist the tendency to remove a functional relationship from the model because we cannot find data with which it can be quantified. We almost always will commit a more serious error by removing an important relationship from the model than we will by guessing at, and subsequently adjusting, the manner in which we represent the relationship in the model.

P21. Reliance on "automated" model parameterization (fascination with methodology): We should resist the tendency to rely on automated ("canned" or "black box") methods to parameterize model relationships. We face a similar temptation during model evaluation (*P30*). One the one hand, we risk using an inappropriate methodology simply because it is convenient to do so and the methodology seems "almost right." On the other hand, we risk becoming overly fascinated with automated parameterization methodologies per se, which can lead to a search for more elegant mathematical and statistical methodologies even though there is no objective demonstration that they are needed.

P22. Use of overly-sophisticated equations: We should not use equations that are more sophisticated or more complicated than is necessary, simply because they are more elegant. This is closely related to the fascination with methodology per se (*P21*). We risk becoming distracted from our modeling objectives by a search for more elegant mathematical representations of system processes, even though there is no objective demonstration that they are needed. An additional risk is that the mathematical sophistication of an equation will imply an unwarranted confidence in our understanding of the relationship.

P23. Underestimation of the importance of negative feedbacks and time lags: We should consider carefully our representation of the negative feedbacks and time lags in our model equations. Negative feedbacks and time lags are almost universal characteristics of complex systems, but they are easily overlooked because cause and effect in complex systems are not always tightly linked in time and space. The most common problems resulting from inappropriate representation of negative feedbacks and time lags include impossibly high or low values of system components and impossibly short response times.

8.2.5 **Execute baseline simulation**

P24. Careless definition of the baseline conditions: We should take care in defining the baseline conditions of the system, since these conditions often are used as the point of reference for both the model evaluation and model application phases of the modeling process. Without well-conceived baseline conditions, we risk losing focus during model evaluation, particularly during sensitivity analysis. We also risk development of a faulty experimental design for model application, which most likely will lead to inappropriate analysis and interpretation of simulation results.

P25. Reliance on automated solutions to mathematical and programming problems: We should not allow built-in programming safeguards to handle unforeseen inconsistencies in our mathematical formulations or programming code. We risk aberrant model behavior resulting from inappropriate, automated procedures that avoid, rather than fix, the problem. We should provide appropriate mathematical or programming solutions ourselves. An example of a common programming problem is the attempt to divide by zero. Often, the process being represented simply does not occur in nature when the system component in the denominator goes to zero. In this case, we simply could imbed the equation in a logical statement that checks to see if the denominator is zero before attempting the division.

8.3 **Pitfalls during model evaluation**

8.3.1 **Assess the reasonableness of the model structure and the interpretability of functional relationships within the model**

P26. Underestimation of the importance of the qualitative aspects of model evaluation: We should not underestimate the importance of the qualitative aspects of model evaluation. We risk losing the most convincing defense of a dynamic model, that is, the fact that all model components are reasonable and are interpretable in ecological, or other appropriate subject-matter, terms.

P27. Acceptance of conceptually flawed functional relationships: We should not accept functional relationships that we know are flawed, even if we have programmed the model (often via logical statements) to prevent these functions from operating under those circumstances for which they are inadequate. Although we may have avoided the flaw numerically, we will have accepted as being useful a model that contains relationships that we know are conceptually flawed, thus obliging us to defend the model as a purely correlative model. Note that avoiding the execution of flawed functional relationships is fundamentally different from avoiding the execution of mathematically undefined operations (*P25*).

8.3.2 **Examine the correspondence between model behavior and the expected patterns of model behavior**

P28. Acceptance or rejection of surprising model behavior without explanation: We should not accept surprising model behavior until we have determined its cause. All model behavior ultimately is understandable because it is based on the rules that have been written in the form of mathematics and computer code. Surprising model behavior can result from conceptual, logical, or computer coding errors, or can "emerge" from complex, but reasonable, interactions among the model components. If we do not explain the source of surprising behavior, we risk creating doubts regarding credibility of the model.

P29. Inappropriate interpretation of the initial phase of model behavior: We must take care to interpret appropriately the initial phase of model behavior. Often our model will exhibit behavior patterns during the first several time steps that are fundamentally different from subsequent patterns. The differences between early patterns of behavior and subsequent patterns usually result from differences between our estimates of the initial values of state variables and values of the state variables that are generated by, or "in step" with, the functional relationships in the model. These differences may represent a "start-up" problem which requires a reinitialization of the model, or, depending on model objectives, may represent precisely the behavior of most interest, as the system responds to new circumstances that we want to explore.

8.3.3 **Examine the correspondence between model projections and the data from the real system**

P30. Reliance on "automated' model evaluation (fascination with methodology): We should resist the tendency to rely on automated ("canned" or "black box") methods to evaluate our model. We face a similar temptation during parameterization of the model (*P21*). We risk deceiving ourselves into thinking that there is a rigorous mathematical and statistical protocol for evaluation of the model. We also risk, inadvertently, allowing our fascination with the methodological details to distract our attention from a thorough evaluation of the real usefulness of the model.

P31. Tendency to believe all of the data from the real system are correct: We should resist the tendency to believe that all data from the real system represent an absolutely reliable picture of the real dynamics of the system; error-free data are extremely rare. We risk underestimating usefulness of the model by interpreting all differences between model projections and real-system data as resulting from inadequacies in the model. Often, these differences result from measurement errors, or other "noise," in the data. In fact, arguably, in some cases, a simulated data point may be a better estimate than a data point from the real system.

P32. Inappropriate interpretation of the statistical tests used in model evaluation: We should not automatically assign importance to all statistically significant differences between model projections and real-system data. We risk underestimating usefulness of the model based on differences that may be of no practical significance. Statistical significance and practical significance are not necessarily synonymous. We encounter an analogous pitfall during the application phase of the modeling process (*P37*).

8.3.4 Determine the sensitivity of the model projections to changes in the values of important parameters

P33. Careless design of the sensitivity analysis: We should design our sensitivity analysis to meet specific objectives within the overall context of the modeling project. We risk generating an enormous amount of useless information by attempting to conduct a "complete" sensitivity analysis involving many levels of each model parameter. We also risk generating misleading information by taking a "shotgun" approach to selection of the parameters to be included in the sensitivity analysis. A "shotgun" selection approach can inadvertently bias our view of the relative sensitivities of model behavior to different parameters. Also, by varying parameter values over inappropriate ranges, we may inadvertently measure the response of the model to situations that are impossible to encounter in the real system. We encounter an analogous pitfall during development of the experimental design in the model application phase of the modeling process (*P35*).

P34. Tendency to equate model sensitivity with model inadequacy: We should resist the tendency to equate overall sensitivity of model behavior to changes in parameter values as a measure of the adequacy or inadequacy of the model, per se. We should interpret sensitivity analysis results on a parameter-specific basis within the context of model objectives. We risk underestimating usefulness of the model as a tool for quantifying our uncertainty about dynamics of the real-system, and as a guide for prioritizing data needs. We also risk misinterpreting results of the simulations that will be run during the model application phase of the modeling process.

8.4 Pitfalls during model application
8.4.1 Develop and execute the experimental design for simulations

P35. Careless experimental design for the model application: We should develop the experimental design for our model application simulations carefully to meet specific project objectives. We must resist the tendency to simulate many scenarios just because we can do so easily. We risk generating useless, and potentially misleading, information by careless selection of management scenarios and/or environmental scenarios that we will simulate. We encounter an analogous pitfall during development of the design for our sensitivity analysis during the model evaluation phase of the modeling process (*P33*).

P36. Tendency to overestimate the range of model applicability: We must be careful to not apply the model to situations in which we violate basic assumptions upon which the model is based. We risk generating meaningless or misleading results that may lead us astray during the development of subsequent models.

8.4.2 Analyze and interpret simulation results

P37. Inappropriate interpretation of statistical tests used in model application: We should not automatically assign importance to all statistically significant differences between or among simulated scenarios. We risk focusing attention on some differences that are of no practical significance. Statistical significance and practical significance are not necessarily synonymous. We encounter an analogous pitfall during the model evaluation phase of the modeling process (*P32*).

8.4.3 Communicate the simulation results

P38. Failure to communicate the numerical results directly in ecological terms: We should take care to interpret all numerical results of our model application simulations within the appropriate ecological (or other appropriate subject-matter) terms whenever possible. We risk reducing usefulness of our simulation results by describing them in a predominantly numerical context without a sufficient number of explicit links to the ecological context within which the numbers can be related directly to our initial questions.

P39. Tendency to imply false expectations regarding the model projections: We should take care to not lose sight of the fact that simulation results are numerical consequences of following logical and mathematical instructions which, collectively, are the model. We risk implying that simulated data are not different from data based on direct physical measurements in the real world, and that the model has a permanence akin to that of a physical entity in the real world. We also risk implying that we can predict the future, which is theoretically impossible except for completely closed systems, which do not exist in the real world.

A candid view of the modeling process

Chapter outline

9.1 What really happens?

In this chapter, we take a candid look at the practical application of simulation modeling. We suggest a strategy for model development that we have found helpful in traversing the pitfall-filled modeling landscape we depicted in Chapter 8. Although theoretically it is convenient to describe the modeling process as proceeding smoothly through the four phases described in Chapters 3–7, in practice we usually cycle through each of these phases several times. We seldom quantify the entire conceptual model before running simulations and evaluating model behavior. Rather, we usually construct a simple "running" model as quickly as possible and then expand it gradually through a series of small additions, each of which we carefully evaluate, until we have quantified the entire model (Fig. 9.1).

Thus, in practice, we engage ourselves in three general types of activities. (1) We develop a preliminary conceptual model of the entire system-of-interest, as well as a

Theory		Pitfall	Practice				
Phase	Step	Number	CM	IDM_1	IDM_i	IDM_{last}	FM
I	a	1–3	√	√	. . .	√	
	b	4,5	√	√	. . .	√	
	c	6	√	√	. . .	√	
	d	7,8	√	√	. . .	√	
	e	9	√	√	. . .	√	
	f	10	√	√	. . .	√	
II	a	11, 12		√	. . .	√	
	b	13		√	. . .	√	
	c	14–16		√	. . .	√	
	d	17–23		√	. . .	√	
	e	24,25		√	. . .	√	
III	a	26, 27		√	. . .	√	
	b	28, 29		√	. . .	√	
	c	30–32		√	. . .	√	
	d	33, 34				√	
IV	a	35, 36					√
	b	37					√
	c	38, 39					√

FIG. 9.1

An integrated view of the ecological modeling process, linking theory (Chapters 3–7), pitfalls (Chapter 8), and practice (this chapter). The checkmarks indicate the theoretical phases corresponding to the indicated stage of practical model development. The *shaded areas* represent the theoretical phases that do not occur during the indicated practical stage.
The theoretical phases of model development. Phase I: Conceptual model formulation (Chapter 4, Fig. 4.1), Phase II: Quantitative model specification (Chapter 5, Fig. 5.1), Phase III: Model evaluation (Chapter 6, Fig. 6.1), Phase IV: Model application (Chapter 7, Fig. 7.1).
The practical stages of model development. *CM*: Conceptual model; IDM_1...last: the series of intermediate developmental models; IDM_{last}: the last intermediate developmental model; *FM*: the final model.

"plan of attack" for quantifying the model piece-by-piece. (2) We construct a series of intermediate developmental models, following our plan of attack, starting with what usually seems like a trivially small portion of the overall model. We quantify, "run," and evaluate each model, and make any necessary adjustments, before proceeding to the next one. We add only a tiny piece to each new model, following our plan, which we also may modify as we proceed. We repeat, quantify, run, evaluate, and adjust sequence until we have reassembled the entire model, which seldom is identical to our original conceptual model. (3) We use the final, "reconstructed," model to address our questions, which we also may have modified as we sharpened our focus during model development.

This is an admittedly "ugly," seldom documented procedure, but each of the practical activities can be related directly to the theoretical phases described in Chapters 3–7. The iterative nature of the approach forces us to constantly re-evaluate our model, both conceptually and quantitatively. This constant re-evaluation sharpens our focus on the project objectives, greatly reduces the likelihood that the model code contains mathematical errors or logical errors, and, invariably, provides additional insight into the dynamics of the system-of-interest. The latter is, without a doubt, one of the most important, although perhaps most underappreciated, benefits of the modeling process.

In the sections that follow, we first relate the three activities that form our practical strategy for model development to the corresponding theoretical phases described in Chapters 3–7 (Fig. 9.1). We then provide three "guided" modeling exercises, which demonstrate the use of this strategy.

9.2 **Preliminary conceptual model (*CM*)**

The objective of the preliminary conceptual model is to qualitatively represent all relevant aspects of the system-of-interest as we currently understand them. This activity follows the first theoretical phase (Ia through If) relatively closely, however, we perform steps Ib through Ie simultaneously as we draw and redraw the box-and-arrow diagram that represents the system-of-interest (Fig. 9.1). Most commonly, this activity is done with paper and pencil (and eraser!), rather than on a computer.

9.2.1 **How to begin**

We may have difficulty deciding how to begin developing the conceptual model. Theoretically, the system has no beginning or end, just connections among components, so we might start anywhere. However, a good rule of thumb is to begin thinking about the model in terms of the type of material that we want to represent and where that material accumulates. That is, we begin thinking about the state variables. Often there is a natural flow of material passing through, or circulating within, the system that provides a "backbone" (a chain of state variables) for the subsequent model structure. Energy flow through an ecosystem, with accumulation points of interest being the energy contained in plants, herbivores, and carnivores; and nutrient cycling within an ecosystem, with accumulation points of interest being the nutrients contained in plants, herbivores, carnivores, and abiotic components, are examples. But we might also think of the individuals in an animal population as "flowing" through the various age classes, with accumulation points of interest being the number of animals in each age class.

Having identified the state variables, this leads us quite naturally to think about the routes by which material moves into, out of, and among the state variables. That is, we begin thinking about material transfers. Identification of the material transfers, in turn, quite naturally focuses our attention on the information that is needed to control rates of material flow. That is, we begin thinking about information transfers to the material transfers which arrive either directly or indirectly from other model components; state variables, driving variables, auxiliary variables, and/or constants. If we immediately think of some components that control a material transfer, we might add them to the model at the time we add the material transfer. But if we find ourselves pondering at length about the various factors that might control any given material transfer, usually it is better to continue with identification of the state variables and material transfers, and postpone our contemplation of the more difficult aspects of model structure until we have identified all of the easier components.

9.2.2 **Adding new components to the model**

If we are in doubt with regard how to represent a new model component, a good rule of thumb is first to try the simplest possible representation: as a constant. If the new component is not constant, try representing it as a driving variable. If the new component is affected by some other component in the system-of-interest, try representing it as an auxiliary variable. If the calculation of sequential values for the new component requires knowledge of its previous value, then represent it as a state variable, with, of course, the appropriate material transfers entering and/or leaving it. To illustrate diagrammatically this sequence of attempting to represent a new model component in various different ways, we will add a new component to our oceanic island animal population model from previous chapters (see Chapter 6, Fig. 6.4A for the most recent conceptual model). We will add "available prey" as the new model component, assuming that available prey affects per capita birth rate of the animal population, first as a constant (Fig. 9.2A), then as a diving variable (Fig. 9.2B), then as an auxiliary variable (Fig. 9.2C), and finally as a state variable (Fig. 9.2D).

9.2.3 **Describing expected patterns**

Once we have a conceptual model that we think adequately represents the system-of-interest, we describe the expected patterns of model behavior (If). This also is a paper-and-pencil activity that typically involves graphing the expected temporal dynamics of key model components, as well as graphically depicting our hypotheses concerning model predictions under different management policies or environmental conditions. These expectations may be based on data from direct observation or experimentation with the system-of-interest, theoretical relationships or generally applicable empirical relationships that are appropriate for the situation being modeled, qualitative information from the literature, or on the opinions of experts.

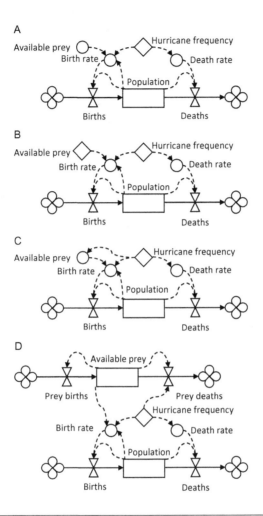

FIG. 9.2

Diagrammatic illustration of the sequence of steps in attempting to represent a new model component (A) first as a constant, (B) then as a driving variable, (C) then as an auxiliary variable, and (D) finally as a state variable. In this illustration, "available prey" is the new model component which is added to the oceanic island animal population model from previous chapters (see Chapter 6, Fig. 6.4A for the previous conceptual model diagram).

9.2.4 Describing the plan of attack

Finally, based on the preliminary conceptual model, we outline a general plan for model development, that is, for quantifying the model. We first identify a tiny subset of the boxes and arrows in the preliminary conceptual model with which we will begin quantification of the model, then we identify a series of increasingly complex intermediate developmental models which we will quantify sequentially en route to obtaining the complete final model. Obviously, for extremely simple models, we hardly need to elaborate a multistep plan for quantitative development of the model. However, such a plan is useful for even relatively simple models, and for more complex models it is essential.

9.3 Intermediate developmental models (IDM_i)

The objective of the series of intermediate developmental models is to obtain an adequate quantitative representation of the preliminary conceptual model. This activity includes all steps in the first (Ia through If), the second (IIa through IIe), and the third (IIIa through IIId) theoretical phases, however, departs from the theory considerably in that we seldom quantify the entire conceptual model before running simulations and evaluating model behavior (Fig. 9.1). Rather, we usually construct a very simple "running" model as quickly as possible and then expand it through a series of tiny changes, following the general plan for model development that we outlined when we finished the preliminary conceptual model (Section 9.1), until the model can be used to address our project objectives. We benefit greatly from having a running model as soon as possible because only then can we see the logical numerical consequences of our assumptions. Often, the puzzle parts that seem to fit together well conceptually do not fit together at all quantitatively. By making each addition to subsequent models as simple as possible, we not only facilitate identification and correction of mathematical errors and logical errors, but we also promote understanding of relationships within the model that control system behavior.

As we proceed through the series of developmental models, our specific objectives (Ia), conceptualization of the system (Ib through Ie), and expected patterns of model behavior (If) likely will change from one model to the next, since we are shifting our focus from the representation of one part of the system to another (Fig. 9.1). Likewise, the manner in which we quantitatively represent any given aspect of the system may change from one developmental model to the next, since the general strategy is to first greatly simplify and then gradually increase complexity. It is unlikely that we will change our choice of the general quantitative structure for the model (IIa), and, as we proceed, it becomes increasingly unlikely that we will change our choice of the basic time unit for simulations (IIb). However, we quite likely will cycle through all the other theoretical steps related to model quantification (IIc through IIe) and evaluation (IIIa through IIId) for each new version of the model that we develop (Fig. 9.1). We also may modify our general plan of attack for model development as we proceed, and may even decide to "fine tune" the objectives of the modeling project as we learn more about the system-of-interest from our interactions with the developing model.

We should proceed through the series of developmental models as quickly as possible. Obviously, "quickly" is a relative term, constrained by overall complexity of the model and the time we have to devote to the modeling project. But the goal is to maintain the continuity of our flow of ideas related to model development.

Thus, even if we feel a strong need to refine our representation of a certain aspect of the model, we should postpone lengthy literature searches and time-consuming data analyses until we have the model fully quantified. That is, until we have reached the last intermediate developmental model (IIId) (Fig. 9.1). Most often we can obtain an adequate first approximation of the needed equations based on rough plots of available data or qualitative descriptions of functional relationships.

When we have fully quantified the last intermediate developmental model (*IDM$_{last}$*) (Fig. 9.1), we then focus on the necessary refinements to our representations of specific aspects of the model. Now is the time that we conduct literature searches and more detailed data analyses to confirm the functional forms of selected model equations and to improve estimates of specific parameters. Having completed these final adjustments, we conduct our last, formal, evaluation of the model, which almost always includes a sensitivity analysis (we comment further on sensitivity analysis in Section 9.3.2). Assuming we fail to reject this last intermediate developmental model as being useless, it becomes the final model that we will use to address the objectives of the modeling project.

9.3.1 The evaluate-adjust cycle for each of the developmental models

We now take a closer look at the manner in which we evaluate and adjust each of the developmental models before proceeding to the next. If we are dissatisfied with the current representation of any given developmental model because it fails to meet one of our first three evaluation criteria (IIIa through IIIc), we return to an earlier theoretical step in model development to make appropriate modifications. The step to which we return depends on why we are dissatisfied (Fig. 9.3). If we are dissatisfied with the reasonableness of model structure or the interpretability of functional relationships in the model, we must return to adjust the conceptual model. If we are dissatisfied with the correspondence between model behavior and our expected patterns of model behavior, we may need to return to adjust the conceptual model, but more likely we will return to adjust functional forms of equations or parameter values in the quantitative model. If we are dissatisfied with the correspondence between model projections and data from the real system, most likely we will return to adjust functional forms of equations or parameter values in the quantitative model. We usually do not conduct a sensitivity analysis until we have reached the last developmental model.

By far the most common type of adjustment we make during the entire modeling process involves the quantitative "tuning" of the various developmental models. That is, the adjustment of functional forms of equations (IIc) and values of parameters (IId) (Fig. 9.1). This is a legitimate activity during model development, but an activity that must be confined strictly in terms of (1) choosing the parameters or functional forms to be adjusted, (2) limiting the number of parameters or functional forms to be adjusted, and (3) determining when to end the "tuning" activities.

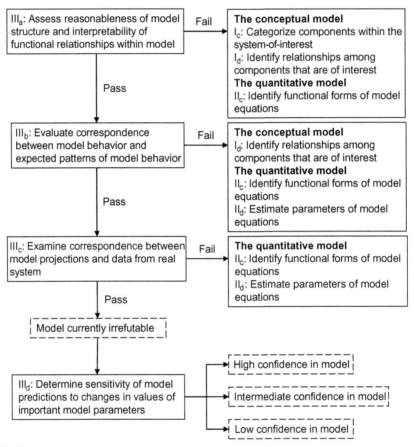

FIG. 9.3

The evaluate and adjust cycle for each of the developmental models. Sensitivity analysis is conducted only on the last intermediate model.

First, discretion must be used in choosing the parameters or functional forms to adjust, as well as the types of adjustments to be made. Adjustments will be made based on our best guesses because no new information external to the model is being drawn upon to make these adjustments, although the nature of the differences between model behavior and our expected patterns of behavior may suggest specific alterations. But only those parameters or functional forms that we have specified in a tentative fashion should be altered, and the type of adjustment made must not result in a parameter value or functional form that can be refuted on conceptual grounds.

Second, we must limit the number of parameter values or functional forms of equations that we adjust in any given model. There is no precise limit, but obviously, as the number of components adjusted increases, so does the number of possible combinations of adjusted values and, hence, the possibility that any improvement in model behavior will result solely from a fortuitous combination of adjustments. It is more common to adjust parameter values than functional forms of equations. This is understandable because the choice of functional forms usually has more profound implications concerning the ecological interpretations of model structure than does the choice of parameter values. We are likely to have based our original choice of general functional forms on a relatively sound understanding of the nature of the relationships in the real system, whereas we are less likely to have based our original choice of specific parameter values on an equally sound basis.

Finally, we should end all tuning of the model before proceeding with formal evaluation (IIIa through IIId) of the last developmental model (Figs. 9.1 and 9.3). If we make statistical comparisons between model projections and real-system data, this clearly marks the point at which tuning activities must be end. If our comparisons between model projections and real-system data are nonstatistical, or we lack evaluation data from the real system, the line is perhaps less clear. But, in good faith, we must end tuning activities before we begin formal evaluation of the last developmental model.

9.3.2 Sensitivity analysis of the last developmental model

This activity follows the corresponding theoretical phase (IIId) quite closely. Our objective is to determine the degree of response, or sensitivity, of model behavior to changes in the various model components. We usually alter the value of one parameter at a time throughout an entire simulation and observe the subsequent effect on model behavior, but we may vary sets of parameters at the same time, or change functional forms of relationships within the model, if these changes seem more appropriate. We usually try to vary any given parameter over a range of values indicative of the degree of the uncertainty associated with our estimation of the parameter.

If we are lucky, model projections will be insensitive to changes in those parameters for which we have poor information. That is, differences among model projections generated with different values of the parameter will have no practical importance within the context of our objectives. If we are not so lucky, and differences in model projections are large enough to be of practical importance, all is not lost. In fact, this information is most useful in (1) identifying important gaps in our knowledge base, and (2) quantifying the uncertainty that we should attribute to our model projections. In such cases, a wise strategy is to make subsequent model projections, those projections that we will use to address project objectives, with several versions of the model which, collectively, encompass our uncertainty.

Thus, by indicating the degree to which uncertainty in our parameter estimates generates uncertainty in model projections, sensitivity analysis provides an appropriate context within which to make, and interpret, our projections with the final model (or the various versions thereof).

9.4 **Final model (*FM*)**

The objective of the final model is to address the objectives of the modeling project. This activity follows the last theoretical phase (IVa through IVc) closely (Fig. 9.1). Because the analysis of simulation results from our initial experimental design almost always raises new questions, before formally reporting our results, we commonly run additional simulations to explore new variants of the most interesting original scenarios. This is analogous to conducting a pilot study in the real world as a basis for future work.

Our simulated pilot study commonly consists of (1) an exploration of the behavior of the system-of-interest under relatively many variants of selected scenarios and, perhaps, (2) a more formal analysis of the most interesting new scenarios. During exploratory simulations, we are interested primarily in identifying general trends in system behavior. If we have a stochastic model, we sometimes conduct these exploratory simulations using a deterministic version of the model in which all random variables are assigned their mean values. If some new scenarios seem worthy of more formal analysis, we could develop and execute a new experimental design, analyze the new results, and incorporate these new findings into our final report.

We formally communicate results of our modeling project (IVc), in general, just as we would formally communicate results of our field or laboratory projects. However, two items are worthy of note. First, since simulated experiments are much less subject to constraints of time and money, it sometimes is difficult to identify a definitive ending point for the modeling project. Indeed, the modeling process ideally is a continuous one, and can help provide continuity in an ongoing program of scientific investigation. Nonetheless, we need to periodically report our results formally.

Second, the standard "methods" and "results" format of technical reports can be awkward for describing the development and application of a simulation model, since the model may be viewed as both a method and as a result. As we mentioned in Chapter 7, we have found that replacing "methods" and "results," if possible, with headings such as "overview of the model," "model description," "model evaluation," and "simulation of [...]", following the format suggested by Grant and Swannack (2008), provides a more comfortable format. We illustrated the use of this format in Chapter 7 (Section 7.4.1).

9.5 **Three guided modeling exercises**

To close this chapter, we provide three "guided" modeling exercises to demonstrate use of the procedure we just described. The first exercise deals with the effect of drought on plant biomass dynamics. The second exercise deals with the allocation of fossil fuel resources to support geothermal energy development. The third exercise deals with the likelihood of survival of an endangered amphibian species. For each exercise our suggestion is to first read the background information and the problem statement. Then, without reading further, develop a simulation model and use it to address the problem, documenting your steps as you go with regard to what you are doing and why you are doing it. Although the examples are relatively simple, try to avoid skipping steps or leaving your reasoning implicit. Then read the rest of the example and compare what you have done with what we did. Note that this is not intended to be a comparison of what you did with the "correct answer." It is intended to be a comparison of what you have done with what some more experienced ecological modelers did, which, hopefully, will be useful in developing your modeling skills. Particularly, your confidence in making the many necessarily arbitrary decisions involved in the modeling process.

9.5.1 Plant biomass dynamics

Background information: There is interest in the effect of drought on the biomass dynamics of a particular plant species which lives in a seasonally varying environment. Growth of this particular species of plant depends not only on current plant size but also on monthly precipitation. Total biomass of this particular species of plant cannot exceed 570 g, due to physiological restrictions. As plant biomass increases, growth rate decreases. When plant biomass is less than 200 g, the plant can grow at the maximum rate possible, however, once biomass reaches 200 g, the growth rate decreases linearly until biomass reaches 570 g, after which the plant cannot grow at all. If monthly precipitation (Table 9.1) is greater than 30 mm plant biomass can increase by a maximum of 100 g per month. If monthly precipitation is less than 15 mm, the plant cannot grow, and if precipitation is between 15 and 30 mm, plant biomass can increase by a maximum of 40 g of biomass per month. (Actually, there is quite a bit of uncertainty associated with the estimate of 15 mm as the lower threshold of precipitation for plant growth.) During December and January, the plant loses 5% of its biomass each month due to winter leaf loss. Finally, at the beginning of January, plant biomass historically has averaged 495 g.

Table 9.1 Monthly precipitation (*precipitation*, mm) in the area where the plant of interest grows.

Month	precipitation
January	14.85
February	28.80
March	33.08
April	30.60
May	14.18
June	14.40
July	15.08
August	28.58
September	34.65
October	31.73
November	27.68
December	13.50

Problem statement: We have been asked to simulate the effects of 2-year droughts of different severities on biomass dynamics of this plant species and assess the ability of the plant to recover during the first postdrought year.

PLEASE DO YOUR "HOMEWORK" BEFORE PROCEDING!

Preliminary conceptual model: Given the problem statement, it seems clear that the material we want to represent is biomass and the most important point of biomass accumulation in our system-of-interest is the plant (Fig. 9.4A). Given the background information, it also seems clear that the routes by which biomass enters and leaves the

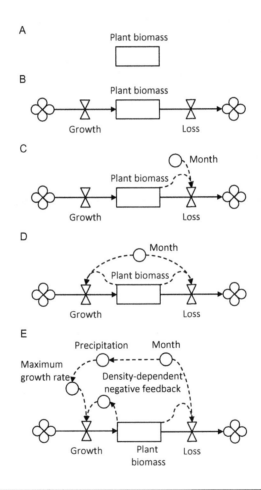

FIG. 9.4

Illustration of the step-by-step development of the conceptual model representing plant biomass dynamics, including representation of (A) the state variable, (B) the material transfers, (C) the driving variable and information transfers needed to calculate biomass loss, (D) additional information transfers needed to calculate biomass growth, (E) auxiliary variables and associated information transfers added to increase ecological interpretability. See text for details.

system-of-interest can be represented simply as biomass growth and biomass loss (Fig. 9.4B). We can represent biomass loss based on information about plant biomass and month of the year (Fig. 9.4C). We also could represent biomass growth based on information about plant biomass and month of the year, leaving the effects on growth of both precipitation and the negative feedback from plant biomass implicit (Fig. 9.4D). However, it seems that explicit representation of the mechanisms by which month affects growth (via the effect of precipitation on maximum potential growth rate) and biomass affects growth (via density-dependent negative feedback) would increase the ecological interpretability of the conceptual model (Fig. 9.4E).

Our description of the expected patterns of model behavior might look something like those sketched in Fig. 9.5. Our reasons for expecting such patterns are based primarily on our prior knowledge of the system-of-interest, as summarized in the background information above, but also on our basic knowledge of plant biology. That is, soil moisture, in this case resulting from precipitation, is a requisite for plant growth, and plant growth form, at least in part, is genetically determined. Thus, if we initialize our simulated plant with 495 g of biomass in January, we would expect biomass to decrease initially due to leaf loss, then increase during the growing season in response to seasonally varying precipitation, with growth rate slowing a bit as biomass approaches 570 g due to genetically imposed constraints, and finally decreasing once again due to winter leaf loss.

Regarding our plan of attack for quantifying our conceptual model, it seems like the driving variable representing precipitation might be a good place to begin (the easiest to quantify), since we just need to input a time series of predetermined values which will be unaffected by our quantification of the rest of the model (Fig. 9.4). The material transfer representing biomass loss looks like the next easiest part to quantify, since it can be represented by a rate constant (0.05, according to the background information, and anticipating that we will chose a monthly time step for our simulations) times plant biomass. Finally, saving what looks like the most

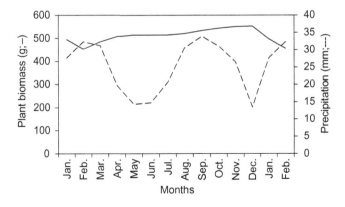

FIG. 9.5

Expected patterns of behavior of the model representing plant biomass dynamics with regard to *precipitation* and *Plant Biomass*. See text for details.

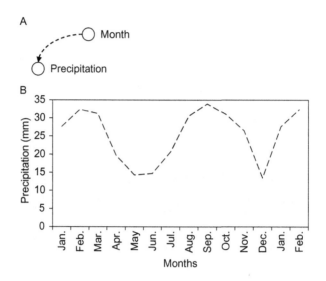

FIG. 9.6

Plant biomass dynamics model. (A) Sketch of the first intermediate developmental model and (B) simulation results.

difficult for last, we will quantify the material transfer representing biomass growth. Growth looks most difficult because its quantification involves representing both the effect of precipitation and the density-dependent effect of plant biomass.

Intermediate developmental models: Following our plan of attack, for our first intermediate developmental model, we will quantify the driving variable representing precipitation (Fig. 9.6A). Quantification involves "putting" (coding) the time series of values representing precipitation contained in Table 9.1, as well as a representation of the 12 months of the year, into our model. For example, we might define *month* = 1 as January, *month* = 2 as February, ..., *month* = 12 as December. Then, as simulated time (*t*) advances month by month, we would calculate

$$Month_{t+1} = Month_t + 1 \tag{9.1}$$

$$\text{if } Month_t > 12 \text{ then } Month_t = 1 \tag{9.2}$$

$$\text{if } Month_t = 1 \text{ then } Precipitation_t = 14.85 \tag{9.3}$$

$$\text{if } Month_t = 2 \text{ then } Precipitation_t = 28.80 \tag{9.4}$$

$$\cdots$$

$$\text{if } Month_t = 12 \text{ then } Precipitation_t = 13.50 \tag{9.5}$$

Comparing simulated precipitation (Fig. 9.6B) with the input data used to parameterize this driving variable (Table 9.1), we verify that our coding for this part of the model is correct.

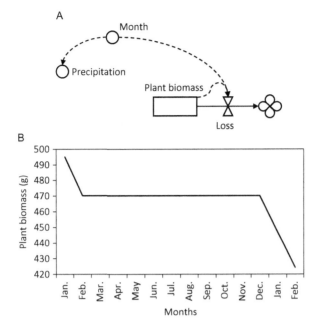

FIG. 9.7

Plant biomass dynamics model. (A) Sketch of the second intermediate developmental model and (B) simulation results.

For our second intermediate developmental model, we will quantify the material transfer representing biomass loss (Fig. 9.7A).

$$Plant\ biomass_0 = 495 \tag{9.6}$$

$$Plant\ biomass_{t+1} = Plant\ biomass_t + (-Loss_t)\ \Delta t \tag{9.7}$$

$$Loss_t = 0.05 \times Plant\ biomass_t\ \text{if}\ Month = 1\ \text{or}\ Month = 12 \tag{9.8}$$

$$Loss_t = 0\ \text{if}\ Month \neq 1\ \text{and}\ Month \neq 12 \tag{9.9}$$

Observing simulated biomass loss from the initial condition of 495 g in January (based on the background information) through the following January (Fig. 9.7B), we conclude that the seasonality of biomass loss seems appropriate (based on the background information).

For our third intermediate developmental model, we will quantify the material transfer representing biomass growth (Fig. 9.8A). Note that we have decided to eliminate temporarily the material transfer representing biomass loss to facilitate the evaluation of biomass growth. Also note that we have decided to modify our original plan of attack a bit, in that we have decided to divide the quantification of biomass growth into two steps. For the first step, shown in Fig. 9.8A, in addition to eliminating biomass loss, we also have decided to eliminate the effect of precipitation on the

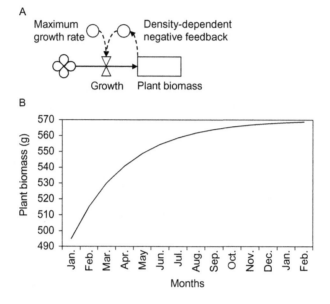

FIG. 9.8

Plant biomass dynamics model. (A) Sketch of the third intermediate developmental model and (B) simulation results.

maximum biomass growth rate, which should facilitate the evaluation of the density-dependent negative feedback of current biomass on growth. Thus, we will represent temporarily the maximum biomass growth rate as a constant equal to 100 g/month, which assumes no limitation due to precipitation.

$$Plant\ biomass_{t+1} = Plant\ biomass_t + (Growth_t)\ \Delta t \qquad (9.10)$$

$$Growth_t = Maximum\ growth\ rate \times Density\text{-}dependent\ negative\ feedback_t \quad (9.11)$$

$$Maximum\ growth\ rate = 100 \qquad (9.12)$$

$$Density\text{-}dependent\ negative\ feedback_t = 1\ if\ Plant\ biomass_t \leq 200 \qquad (9.13)$$

$$Density\text{-}dependent\ negative\ feedback_t = 0\ if\ Plant\ biomass_t \geq 570 \qquad (9.14)$$

$$Density\text{-}dependent\ negative\ feedback_t = 1.5405 - 0.0027 \times Plant\ biomass_t\ if\ 200$$
$$< Plant\ biomass_t < 570 \qquad (9.15)$$

Observing simulated biomass growth from the initial condition of 495 g in January through the following January (Fig. 9.8B), we conclude that the convex form of biomass growth, approaching 570 g asymptotically (ever more slowly) seems appropriate (based on the background information, which describes a biomass growth rate that decreases linearly with increasing plant biomass after a threshold biomass of 200 g is surpassed).

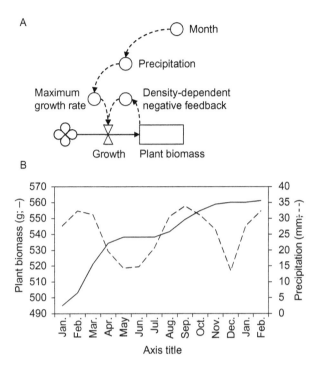

FIG. 9.9

Plant biomass dynamics model. (A) Sketch of the fourth intermediate developmental model and (B) simulation results.

For our fourth intermediate developmental model, we will requantify the material transfer representing biomass growth, now representing maximum growth rate as a function of precipitation, that is, converting maximum growth rate from a constant into an auxiliary variable (Fig. 9.9A).

$$Maximum\ growth\ rate_t = 0\ if\ Precipitation_t < 15 \qquad (9.16)$$

$$Maximum\ growth\ rate_t = 100\ if\ Precipitation_t > 30 \qquad (9.17)$$

$$Maximum\ growth\ rate_t = 40\ if\ 15 \le Precipitation_t \le 30 \qquad (9.18)$$

Observing simulated biomass growth from the initial condition of 495 g in January through the following January (Fig. 9.9B), we conclude that the periods of slower growth and no growth correlated with the seasonally varying precipitation, which are superimposed on the generally convex form of biomass growth, seem appropriate (based on the background information, and our basic knowledge of plant biology).

For our fifth, and last, intermediate developmental model, we will replace the material transfer representing biomass loss, thus arriving at a completely quantified version of our conceptual model (Fig. 9.10A).

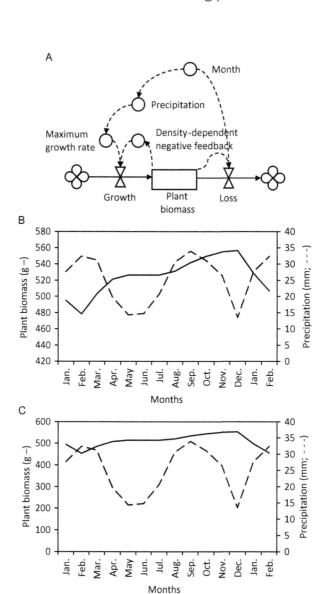

FIG. 9.10

Plant biomass dynamics model. (A) Sketch of the fifth, and last, intermediate developmental model and simulation results (B) before and (C) after "tuning" (calibration).

$$Plant\,biomass_{t+1} = Plant\,biomass_t + (Growth_t - Loss_t)\,\Delta t \qquad (9.19)$$

Observing simulated biomass growth from the initial condition of 495 g in January through the following January (Fig. 9.10B), we conclude that the periods of slower growth and no growth correlated with the seasonally varying precipitation, as well as the seasonality of biomass loss due to winter leaf loss, seem appropriate (based on the background information, and our basic knowledge of plant biology). However, we notice that simulated plant biomass at the beginning of the second January (after 12 months of simulation) is higher than expected (529 g vs 495 g). This difference does not seem large enough to suspect serious problems with model structure, but does suggest that we might adjust ("tune" or calibrate) the parameter value controlling the rate of biomass loss due to winter leaf loss to achieve a better correspondence between model behavior and our expected patterns of model behavior. Indeed, if we increase the value of this parameter from the original estimate of 0.05 to 0.10, the resulting simulated plant biomass at the beginning of the second January is 498 g (Fig. 9.10C).

Having completed the quantification of our model, prior to applying the model to address the problem at hand, we would like to get a feel for how the uncertainty associated with the estimate of 15 mm as the lower threshold of precipitation for plant growth (mentioned in the background information) might influence our confidence in simulation results. To investigate this, we will conduct a sensitivity analysis in which we will first raise the threshold value to 25 mm and then lower it to 5 mm, and monitor the effect of these changes on plant biomass dynamics. Quantification of these changes involves replacing the "15" in Eqs. (9.16) and (9.18) first with a "25" and then with a "5."

Although the annual maximums and minimums of plant biomass were not noticeably sensitive to these threshold changes, there were some noteworthy differences in the midyear patterns of plant growth, with biomass levels in July differing by as much as 18 g (508 g vs 526 g) (Fig. 9.11). Given these results of sensitivity analysis, it seems wise to address our project objective using multiple versions of our final model, which encompass, or explicitly represent, this uncertainty.

Final model: To use our final model to address the objectives of the modeling project, that is, to simulate the effects of 2-year droughts of different severities on biomass dynamics of the plant species and assess the ability of the plant to recover during

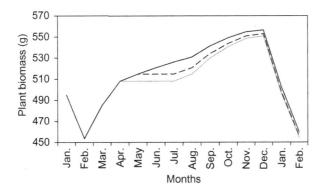

FIG. 9.11

Plant biomass dynamics model. Results of sensitivity analysis comparing simulated plant biomass dynamics assuming the lower annual precipitation threshold for plant growth was 15 mm (baseline, *dashed line*), 25 mm *(gray line)*, and 5 mm *(black line)*.

the first postdrought year, we will simulate a series of increasingly severe droughts. More specifically, our experimental design will consist of a series of 10, 3-year simulations in which we reduce annual precipitation during the first 2 years from its baseline level to 10% of its baseline level, in increments of 10% (100%, 90%, ..., 20%, and 10% of baseline). For all simulations, we will return precipitation to its baseline level during the third year, and we will maintain the baseline seasonal distribution of precipitation throughout all 3 years. To reflect the uncertainty associated with our estimate of the lower annual precipitation threshold for plant growth, we will execute this experimental design using each of three versions of our model in which the lower threshold is (1) 15 mm (baseline), (2) 25 mm, and (3) 5 mm. To facilitate the execution of these simulations, and to visually highlight the experimental design, we will modify slightly our conceptual model by adding two constants (*drought effect* and *lower growth threshold*) and an auxiliary variable (*drought precipitation*) (Fig. 9.12). Quantification of these changes takes the following form:

$$Drought\,precipitation_t = Drought\,effect_t \times Precipitation_t \qquad (9.20)$$

$$Drought\,effect_t = 1 \text{ if } t > 24 \qquad (9.21)$$

$$Drought\,effect_t = Scenario \text{ if } t \le 24 \qquad (9.22)$$

where the value of *scenario* is 1.0, 0.9, ...0.2, or 0.1, depending on the 2-year drought scenario being simulated. We also modify the quantification of *maximum growth rate* by replacing the "15" in Eqs. (9.16) and (9.18) with *lower growth threshold*, where the value of *lower growth threshold* is 15, 25, or 5, depending on the version of the model we are using.

Simulation results indicate that the plant can recover from 2-year droughts severe enough to prevent growth during the entire drought period, regardless of the version of the model used (Fig. 9.13). Simulations using model versions with the lower annual precipitation threshold for growth set at 15 mm (baseline, Fig. 9.13B), 25 mm (Fig. 9.13C), and 5 mm (Fig. 9.13D) indicated that growth was prevented completely

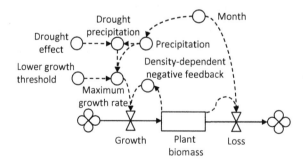

FIG. 9.12

Plant biomass dynamics model. Modified conceptual model highlighting the experimental design for model application simulations (note the addition of "drought effect," "lower growth threshold," and "drought precipitation").

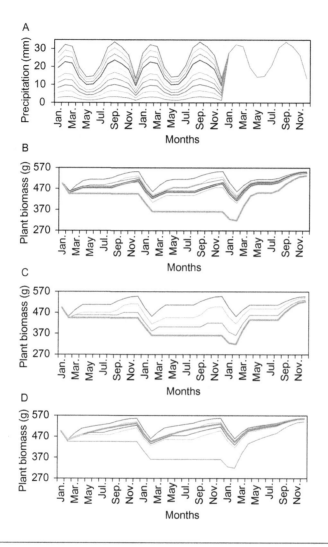

FIG. 9.13

Plant biomass dynamics model. Results of simulations representing (A) increasingly severe droughts and their effects on plant biomass dynamics using model versions with the lower annual precipitation threshold for plant growth set at (B) 15 mm (baseline), (C) 25 mm, and (D) 5 mm. See text for details of experimental design.

during droughts in which precipitation was reduced to 40%, 70%, and 10% of baseline, respectively. However, even with precipitation reduced to 10% of baseline, plant biomass reached peak levels during the first postdrought year that were within 20, 25, and 16 g, respectively, of peak biomass under baseline precipitation (Fig. 9.13B, C, and D, respectively). Under less severe drought conditions, biomass dynamics during the two drought years were affected primarily by postponement of the beginning of the growing season from early spring to later in the year.

These simulation results from our initial experimental design raise some interesting new questions regarding the effects of droughts that are restricted to specific seasons of the year, and also perhaps regarding the possibility that the rate of biomass loss might not remain constant during extended periods without plant growth. We leave it to your imagination and ingenuity to design and conduct simulated pilot studies to explore these questions, or other questions that might occur to you. We also leave you with task of formally communicating the results of this modeling project using the format we illustrated in Chapter 7 (Section 7.4.1).

9.5.2 Geothermal energy development

Background information: There is concern that we will not be able to maintain our current lifestyle based solely on the nonrenewable fossil fuel resources, which currently are our sole source of energy. Currently, we sell to other countries all of the "surplus" useable energy derived from the fossil fuel we extract from the earth. That is, we sell all of the useable energy that is not needed to maintain our current lifestyle. We are considering using some of the surplus useable energy to support the research necessary to develop geothermal energy technology. Recent studies indicate that 100,000 units of fossil fuel remain in the earth. Currently, we can extract 50% of the remaining fossil fuel each year, we can convert 1 unit of fossil fuel into 0.85 units of useable energy, and we need 1500 units of useable energy per year to maintain our current lifestyle. Experts estimate a 5% decrease in the annual fossil fuel extraction rate for every 10,000 units removed from the earth, since what remains is increasingly more difficult to extract. However, experts also estimate that we will be able to increase number of useable energy units we can obtain from each unit of fossil fuel by 0.01 units per year. That is, next year we will be able to convert 1 unit of fossil fuel into 0.86 units of useable energy, the year after that we will be able to convert 1 unit of fossil fuel into 0.87 units of useable energy, and so on. Finally, experts estimate that we will need to invest a total of 5000 units of useable energy into the development of geothermal technology before this new source of energy, which should provide more than enough energy to maintain our current lifestyle, will become operational.

Problem statement: We have been asked to simulate alternative future scenarios to answer the following questions. (1) How long we will be able to maintain our current lifestyle, assuming we continue to extract fossil fuel at the maximum rate possible and that each year we sell all of our surplus useable energy? (2) What is the minimum proportion of our surplus useable energy we could devote to the development of geothermal technology in order for geothermal energy to be operational before we have to lower our lifestyle?

PLEASE DO YOUR "HOMEWORK" BEFORE PROCEDING!

Preliminary conceptual model: Given the background information and the problem statement, it seems clear that one material we want to represent is fossil fuel and an important point of fossil fuel accumulation in our system-of-interest is the fossil fuel remaining in the earth (Fig. 9.14A). Given the background information, it also

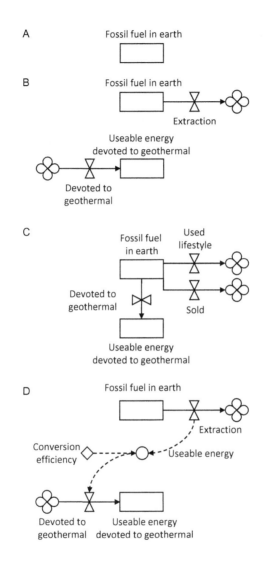

FIG. 9.14

Illustration of the step-by-step development of the conceptual model representing geothermal energy development, including representation of (A) one state variable, (B) a second state variable, a material transfer leaving the first state variable with its destination yet to be defined, and a material transfer entering the second state variable with its origin yet to be defined, (C) an option connecting the two state variables via a material transfer, with two material transfers leaving the system-of-interest, (D) a better option connecting the two state variables via information transfers, with the addition of a driving variable and an auxiliary variable,

(Continued)

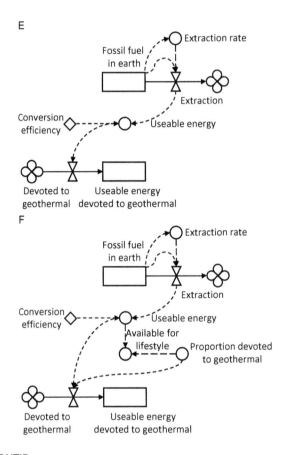

FIG. 9.14—CONT'D

(E) the addition of an auxiliary variable and associated information transfers controlling *extraction*, and (F) the addition of a constant, an auxiliary variable, and the associated information transfers to represent the fact that *useable energy* could be allocated both to supporting current lifestyle and to geothermal development. See text for details.

seems clear that this accumulation of fossil fuel is diminished solely via extraction. However, how to route the extracted material is more puzzling. Given the problem statement, it seems that fossil fuel units in their "raw" form, that is, in their physical and chemical form at the moment of extraction, somehow need to be converted into, and accumulated as, units of useable energy devoted to the development of geothermal technology (Fig. 9.14B). One option might be to represent the fossil fuel units remaining in the earth as useable energy units and connect *Fossil Fuel in Earth* to *Useable Energy Devoted to Geothermal* via material transfers, indicating the losses of useable energy units via material transfers leaving the system-of-interest (Fig. 9.14C). (Remember that state variables connected by material transfers must have the same units of measure.) This option would work if the conversion rate of fossil fuel units to useable energy units was constant. For example, if the conversion rate remained at 0.85 units of useable energy per unit of fossil fuel, we could convert the 100,000 units of fossil fuel remaining in the earth into 85,000 units of useable energy as the initial condition of *Fossil Fuel in Earth*. Since extraction rate is represented as a proportion in the background information, we could just as easily extract a proportion of useable energy units as fossil fuel units. However, since the conversion rate of fossil fuel units to useable energy units changes over time, this option seems awkward at best (we would need to have a material transfer entering *Fossil Fuel in Earth* to account for the increasing number of useable energy units that each unit of fossil fuel was worth). A better option would be to make the connection between the two state variables via information transfers. This frees us to make the appropriate unit conversions en route from *Fossil Fuel in Earth* to *Useable Energy Devoted to Geothermal* via a driving variable (*conversion efficiency*) and an auxiliary variable (*useable energy*), and the associate information transfers (Fig. 9.14D).

Having decided on the option to connect the two state variables via information transfers, we now can focus our attention on the information network as a whole. We note that we still need to represent control of the material transfer *extraction* as a function *Fossil Fuel in Earth*, which we will do by adding an auxiliary variable (*extraction rate*) and the associated information transfers (Fig. 9.14E). Finally, we want to represent the fact that *useable energy* could be allocated both to supporting current lifestyle and to geothermal development (and any "extra" could be sold). One way to do this would be to add a constant (*proportion devoted to geothermal*) and an auxiliary variable (*available for lifestyle*) and the associated information transfers (Fig. 9.14F). Note that we have chosen not to represent energy sales explicitly, since our interest is only in *available for lifestyle* given different values for *proportion devoted to geothermal*.

Our description of the expected patterns of model behavior, based on the background information above, might look something like those sketched in Fig. 9.15. Assuming we extract fossil fuel at the maximum rate, we would expect essentially exponential decreases in *Fossil Fuel in Earth* (Fig. 9.15A), *extraction rate* (Fig. 9.15B), and *useable energy* (Fig. 9.15C). Assuming we held *proportion devoted to geothermal* constant, we would expect *Useable Energy Devoted to Geothermal* to increase at a decreasing rate, with the rate of increase being higher, of course, the larger the *proportion devoted to geothermal* (Fig. 9.15D).

Regarding our plan of attack for quantifying our conceptual model, it seems like the driving variable *conversion efficiency* might be a good place to begin (Fig. 9.14), since we just need to input a time series of predetermined values (Table 9.2) which will be unaffected by our quantification of the rest of the model. The material transfer *extraction* looks like the next easiest part to quantify, since it requires only the quantification of *extraction rate*. Next, we will quantify the auxiliary variables *useable energy* and *available for lifestyle*, which also will require quantification of the constant *proportion devoted to geothermal*. Finally, we will quantify the material transfer *devoted to geothermal*.

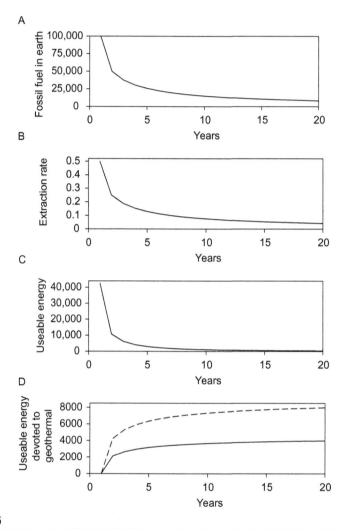

FIG. 9.15

Expected patterns of behavior of the model representing geothermal energy development with regard to (A) *Fossil Fuel in Earth*, (B) *extraction rate*, (C) *useable energy*, and (D) *Useable Energy Devoted to Geothermal*. See text for details.

Intermediate developmental models: Following our plan of attack, for our first intermediate developmental model, we will quantify the driving variable *conversion efficiency* (Fig. 9.16A). Quantification involves coding the time series of values representing *conversion efficiency* contained in Table 9.2. For example, as simulated time (*t*) advances year by year, we would calculate

$$\text{if } t = 1 \text{ then } Conversion\ efficiency_t = 0.85 \qquad (9.23)$$

Table 9.2 Efficiency of converting fossil fuel units into useable energy units (*conversion efficiency*).

Year of simulation	conversion efficiency
1	0.85
2	0.86
3	0.87
...	...
18	1.02
19	1.03
20	1.04

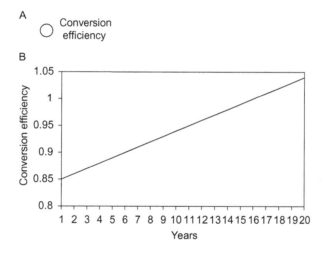

FIG. 9.16

Geothermal energy development model. (A) Sketch of the first intermediate developmental model and (B) simulation results.

$$\text{if } t = 2 \text{ then } Conversion\ efficiency_t = 0.86 \qquad (9.24)$$

$$...$$

$$\text{if } t = 20 \text{ then } Conversion\ efficiency_t = 1.04 \qquad (9.25)$$

Comparing simulated *conversion efficiency* (Fig. 9.16B) with the input data used to parameterize this driving variable (Table 9.2), we verify that our coding for this part of the model is correct.

For our second intermediate developmental model, we will quantify the auxiliary variable *extraction rate* and the material transfer *extraction* (Fig. 9.17A).

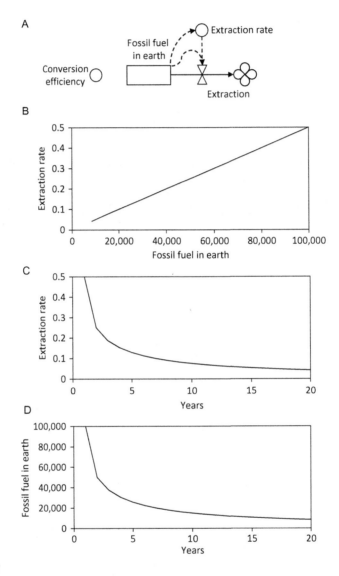

FIG. 9.17

Geothermal energy development model. (A) Sketch of the second intermediate developmental model, (B) simulated *extraction rate* as a function of simulated *Fossil Fuel in Earth*, and simulated decreases over time in (C) *extraction rate* and (D) *Fossil Fuel in Earth*.

$$Fossil\,fuel\,in\,earth_0 = 100,000 \tag{9.26}$$

$$Fossil\,fuel\,in\,earth_{t+1} = Fossil\,fuel\,in\,earth_t + (-Extraction_t)\,\Delta t \tag{9.27}$$

$$Extraction_t = Extraction\,rate_t \times Fossil\,fuel\,in\,earth_t \tag{9.28}$$

$$Extraction\,rate_t = 0.000005 \times Fossil\,fuel\,in\,earth_t \tag{9.29}$$

Observing simulated *extraction rate* as a function of simulated *Fossil Fuel in Earth* (Fig. 9.17B), we verify that our coding for this part of the model is correct. Also, the simulated decreases over time in *extraction rate* (Fig. 9.17C) and *Fossil Fuel in Earth* (Fig. 9.17D) seem appropriate based on our expected patterns of model behavior (Fig. 9.15B and A, respectively).

For our third intermediate developmental model, we will quantify the auxiliary variables *useable energy* and *available for lifestyle*, which also will require quantification of the constant *proportion devoted to geothermal* (Fig. 9.18A).

$$Useable\,energy_t = Conversion\,efficiency_t \times Extraction_t \tag{9.30}$$

$$Available\,for\,lifestyle_t = (1 - Proportion\,devoted\,to\,geothermal) \times Useable\,energy_t \tag{9.31}$$

where *proportion devoted to geothermal* depends on the scenario being simulated.

Simulated decreases over time in *useable energy* (Fig. 9.18B) seem appropriate based on our expected patterns of model behavior (Fig. 9.15C), and simulated decreases in *available for lifestyle* seem appropriately related to changes in *proportion devoted to geothermal* (Fig. 9.18C).

For our fourth, and last, intermediate developmental model, we will quantify the material transfer *devoted to geothermal* (Fig. 9.19A).

$$\begin{aligned}Useable\,energy\,devoted\,to\,geothermal_{t+1}\\ = Useable\,energy\,devoted\,to\,geothermal_t + (Devoted\,to\,geothermal_t)\,\Delta t \tag{9.32}\end{aligned}$$

$$Devoted\,to\,geothermal_t = Useable\,energy_t \times Proportion\,devoted\,to\,geothermal \tag{9.33}$$

Simulated increases over time in *Useable Energy Devoted to Geothermal* (Fig. 9.19B) seem appropriately related to changes in *proportion devoted to geothermal*, and also seem appropriate based on our expected patterns of model behavior (Fig. 9.15D).

Having completed the quantification of our model, prior to applying the model to address the problem at hand, we would like to get a feel for how the uncertainty associated with two estimates mentioned in the background information might influence our confidence in simulation results for *useable energy*. Specifically, we wonder how uncertainty associated with the estimates of (1) a 5% annual decrease in *extraction rate* and (2) an annual increase in *conversion efficiency* of 0.01 units might

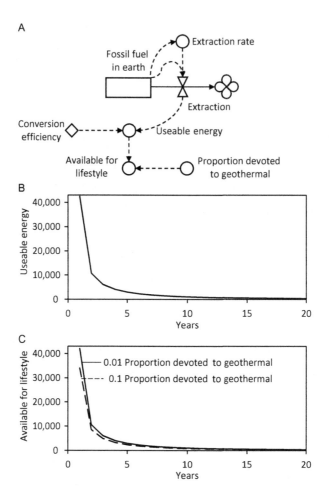

FIG. 9.18

Geothermal energy development model. (A) Sketch of the third intermediate developmental model, (B) simulated decreases over time in *useable energy* and (C) simulated decreases over time in *available for lifestyle* with *proportion devoted to geothermal* equal to 0.01 *(black line)* and 0.1 *(dashed line)*.

affect our estimate of the rate of decline in *useable energy*. To investigate this, we will conduct a sensitivity analysis consisting of 10 simulations in which we (1) increase the value of the coefficient in Eq. (9.29) from 0.000005 to 0.000009 in increments of 0.000001 (thus increasing the annual rate of decrease in *extraction rate*) and (2) increase the annual *conversion efficiency* from 0.01 to 0.05 in increments of 0.01.

Results of sensitivity analysis indicate that (1) the simulated decreases in *extraction rate* had a pronounced effect on model projections of *useable energy* (Fig. 9.20A), whereas the simulated increases in *conversion efficiency* had a negligible

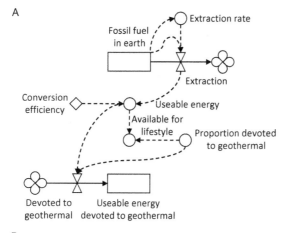

A

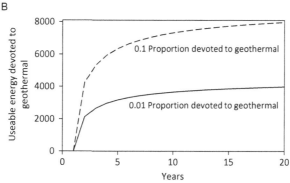

B

FIG. 9.19

Geothermal energy development model. (A) Sketch of the fourth, and last, intermediate developmental model, and (B) simulated increases over time in *Useable Energy Devoted to Geothermal* with *proportion devoted to geothermal* equal to 0.01 *(black line)* and 0.1 *(dashed line)*.

effect (Fig. 9.20B). The amount of *useable energy* during year 2 of simulated time decreased from 10750 units to 774 units as the value of the coefficient in Eq. (9.29) was increased from 0.000005 to 0.000009, whereas the amount of *useable energy* increased from 10,750 units to only 11,250 units as the annual increase in *conversion efficiency* increased from 0.01 to 0.05. Given these results of sensitivity analysis, it seems wise to address our project objective using multiple versions of our final model, which encompass, or explicitly represent, the uncertainty associated with the estimate of *extraction rate*.

Final model: To address the first objective of the modeling project, that is, to estimate how long we will be able to maintain our current lifestyle, assuming we continue to extract fossil fuel at the maximum rate possible, we will use the five versions of our model we used to examine the sensitivity of *useable energy* to changes in

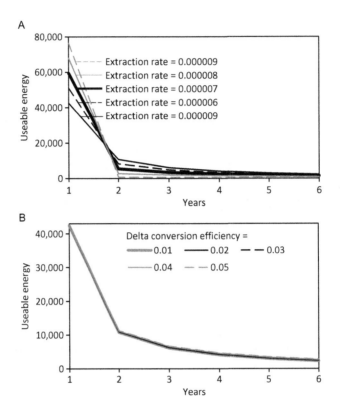

FIG. 9.20

Geothermal energy development model. Results of sensitivity analysis comparing simulated decreases in *useable energy* with (A) *extraction rate* equal to (0.000005), (0.000006). (0.000007), (0.000008), and (0.000009), and (B) with *conversion efficiency* equal to (0.01), (0.02). (0.03), (0.04), and (0.05).

the rate of decrease in *extraction rate*. Our experimental design will consist of five simulations, one with each version of the model, in which we monitor *useable energy* and note the year of simulated time when it first becomes less than the 1500 units needed to maintain our current lifestyle. Actually, these are exactly the simulations we ran during our sensitivity analysis, however, the context within which we will interpret the simulation results has changed. Within the context of addressing our first objective, we estimate that we will be able to maintain our current lifestyle based on *useable energy* extracted from fossil fuel for as many as 8 years or as few as 2 years, depending on our estimate of the rate of decrease in *extraction rate* (Fig. 9.21A).

To address the second objective of the modeling project, that is, to estimate the minimum proportion of *useable energy* we could devote to the development of geothermal technology in order for geothermal energy to be operational before we have to lower our lifestyle, we will use the same five versions of our model.

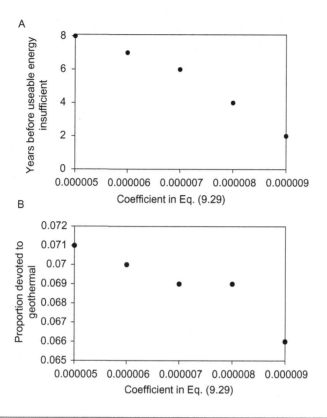

FIG. 9.21

Geothermal energy development model. Results of simulations estimating (A) how many years we will be able to maintain our current lifestyle based on *useable energy* extracted from fossil fuel, and (B) the minimum *proportion devoted to geothermal* for geothermal energy to be operational before we have to lower our lifestyle. See text for details of experimental design.

Our experimental design will consist of five series of simulations, one with each version of the model, in which sequentially increase *proportion devoted to geothermal* from 0 until we reach the value that results in an amount of *Useable Energy Devoted to Geothermal* greater than the 5000 units required to complete technological development the year that the value of *available for lifestyle* first becomes less than the 1500 units needed to maintain our current lifestyle. Simulation results suggest that the minimum *proportion devoted to geothermal* for geothermal energy to be operational before we have to lower our lifestyle ranges from 0.071 to 0.066 depending on our estimate of the rate of decrease in *extraction rate* (Fig. 9.21B).

As with our plant biomass example, these simulation results from our initial experimental design raise some interesting new questions, which we leave to your imagination and ingenuity to identify and explore. We also leave you, as before, with task of formally communicating the results of this modeling project.

9.5.3 **Amphibian species survival**

Background information: There is concern about the future of an endangered amphibian species. The aquatic (larval) and adult stages of the species have been studied fairly well, but the juvenile stage is hard to observe in the field, and, hence, there is considerable debate regarding juvenile survival. The species breeds once per year. The offspring spend 1 year each as larvae (tadpoles) and juveniles, reaching sexual maturity at the beginning of their third year of life. Annual survival is stage-specific and is 5% and 20% for larvae and adults, respectively. Estimates of juvenile survival vary widely from year to year, with estimates ranging from 0.5% to 3%, with 2.5% being the most recent estimate. The maximum birth rate has been estimated at 3000 eggs per female per year (sex ratio 1M:1F). Realized birth rate is dependent on the number of adults currently in the population. When the number of adults is essentially 0, realized natality is maximum, and when the number of adults reaches 300, no breeding occurs. Over the past 10 years, the number of adults in the population has fluctuated between 80 and 280. Currently, the best estimates suggest that there are 102 adults in the population.

Problem statement: We have been asked to develop a population projection model for this species that explicitly represents the uncertainty associated with juvenile survival, and provide estimates of the likelihood that the species will survive. PLEASE DO YOUR "HOMEWORK" BEFORE PROCEDING!

Preliminary conceptual model: Given the background information and the problem statement, it seems clear that the material we want to represent is numbers of individuals. However, not all individuals are the same with regard to some important attributes that affect population dynamics. Both birth rates and survival rates depend on the life stage of individuals. Thus, we cannot represent all individuals in a single state variable because we will lose the ability to know how many individuals are in each life stage. But we could represent the development of cohorts of individuals through larval, juvenile, and adult life stages in a "main chain" of state variables (see Chapter 2, Section 2.7.1), to which we could add the appropriate material transfers representing birth and stage-specific mortalities (Fig. 9.22A). We then could add the stage-specific constants and associated information transfers needed to calculate birth, survival, and mortality material transfers (Fig. 9.22B). Finally, we could add the auxiliary variable and associated information transfers needed to calculate the density-dependent negative feedback of number of adults on number of births (Fig. 9.22C). We could add an auxiliary variable representing total population size, but since our interest is in estimating the likelihood of species survival, and larvae will become adults in 3 years, it seems that monitoring number of adults should be sufficient (if there are no adults for 3 consecutive years, the species will have become extinct). Also, the field data available for model evaluation include only the estimated number of adults.

Our description of the expected pattern of model behavior might look something like that sketched in Fig. 9.23. Our reasons for expecting this pattern are based primarily on the background information above, but also on our knowledge of the

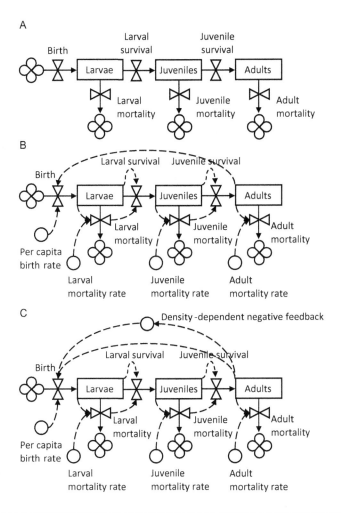

FIG. 9.22

Illustration of the step-by-step development of the conceptual model representing amphibian population dynamics, including representation of (A) the "main chain" of state variables and associated material transfers, (B) the constants and associated information transfers needed to calculate birth, survival, and mortality, and (C) the auxiliary variable and associated information transfers needed to calculate the density-dependent negative feedback on birth. See text for details.

principles of population growth, as well as the fact that we probably will initialize our model with zero larvae, zero juveniles, and 102 adults (since we do not have field estimates of the numbers of larvae and juveniles). Based on the fact that there is density-dependent negative feedback on births, we would expect the population to exhibit an essentially logistic (S-shaped) growth form as it approached the

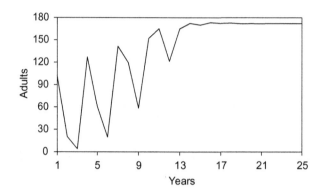

FIG. 9.23

Expected pattern of behavior of the model representing amphibian species survival with regard to *Adults*. See text for details.

environmental carrying capacity. If we initialize the model with zero larvae and juveniles, we would expect the population to fluctuate about this generally logistic growth curve as successive cohorts of different sizes entered the population, until the population reached carrying capacity and established the stable stage distribution (in which the proportions of individuals in each stage class no longer changed). Thereafter, the population would be in equilibrium.

Regarding our plan of attack for quantifying our conceptual model, we will begin with the "main chain" of state variables and the associated material transfers depicted in Fig. 9.22A. We will "send" a single cohort of individuals through the "main chain" and, assuming all survive, verify the 3-year time lag from *birth* to *adults*. Next we will quantify the stage-specific constants needed to calculate *birth* and the various survival and mortality material transfers (Fig. 9.22B). Finally, we will quantify the auxiliary variable *density-dependent negative feedback* and adjust *birth* accordingly (Fig. 9.22C).

Intermediate developmental models: Following our plan of attack, for our first intermediate developmental model, we will "send" a single cohort of 100 individuals through the "main chain" of state variables (Fig. 9.22A), assuming all survive.

$$Larvae_0 = 0 \tag{9.34}$$

$$Juveniles_0 = 0 \tag{9.35}$$

$$Adults_0 = 0 \tag{9.36}$$

$$Larvae_{t+1} = Larvae_t + (Birth_t - Larval\ mortality_t - Larval\ survival_t)\ \Delta t \tag{9.37}$$

$$Juveniles_{t+1} = Juveniles_t + (Larval\ survival_t - Juvenile\ mortality_t - Juvenile\ survival_t)\ \Delta t \tag{9.38}$$

$$Adults_{t+1} = Adults_t + (Juvenile\ survival_t - Adult\ mortality_t)\ \Delta t \qquad (9.39)$$

$$Birth_t = 100\ \text{if}\ t = 0 \qquad (9.40)$$

$$Birth_t = 0\ \text{if}\ t > 0 \qquad (9.41)$$

$$Larval\ mortality_t = 0 \qquad (9.42)$$

$$Juvenile\ mortality_t = 0 \qquad (9.43)$$

$$Adult\ mortality_t = 0 \qquad (9.44)$$

$$Larval\ survival_t = Larvae_t \qquad (9.45)$$

$$Juvenile\ survival_t = Juveniles_t \qquad (9.46)$$

Observing the time lags between simulated *birth*, *Larvae*, *Juveniles*, and *Adults*, we verify that the model code generates the appropriate 3-year time lag from *birth* to *adults* for a cohort of individuals (Fig. 9.24).

For our second intermediate developmental model, we will quantify the stage-specific constants needed to calculate *birth* and the various survival and mortality material transfers (Fig. 9.22B).

$$Per\ capita\ birth\ rate = 3000 \qquad (9.47)$$

$$Larval\ mortality\ rate = 1 - 0.05 \qquad (9.48)$$

$$Juvenile\ mortality\ rate = 1 - 0.025 \qquad (9.49)$$

$$Adult\ mortality\ rate = 1 - 0.2 \qquad (9.50)$$

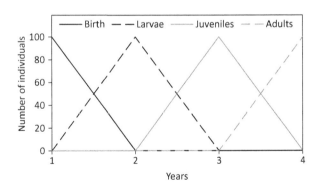

FIG. 9.24

Amphibian species survival model. Simulation results of the first intermediate developmental model (see Fig. 9.22A for a sketch of the associated conceptual model). Lines represent birth *(black line)*, Larvae *(black dashed line)*, Juveniles *(gray line)*, and Adults *(gray dashed line)*.

Note that we have used the most recent estimate of juvenile survival (2.5%). We now update the material transfer equations and the initial conditions:

$$Birth_t = (Adults_t/2) \times Per\ capita\ birth\ rate \tag{9.51}$$

$$Larval\ mortality_t = Larvae_t \times Larval\ mortality\ rate \tag{9.52}$$

$$Juvenile\ mortality_t = Juveniles_t \times Juvenile\ mortality\ rate \tag{9.53}$$

$$Adult\ mortality_t = Adults_t \times Adult\ mortality\ rate \tag{9.54}$$

$$Larval\ survival_t = Larvae_t - Larval\ mortality_t \tag{9.55}$$

$$Juvenile\ survival_t = Juveniles_t - Juvenile\ mortality_t \tag{9.56}$$

$$Adults_0 = 102 \tag{9.57}$$

Since we have not yet quantified *density-dependent negative feedback*, we would expect our second intermediate developmental model to generate essentially exponential population growth. Since we are initializing the population with all adults, we would expect fluctuations about this generally exponential growth curve until the population established the stable stage distribution, after which the population would exhibit a constant annual rate of exponential growth (λ). Observing simulated population growth, we note the model generates this expected pattern (Fig. 9.25).

For our third, and last, intermediate developmental model, we will quantify the auxiliary variable *density-dependent negative feedback*, assuming a linear relationship, and adjust *birth* accordingly (Fig. 9.22C).

$$Density\text{-}dependent\ negative\ feedback_t = 1 - 0.00333 \times Adults_t \tag{9.58}$$

$$Birth_t = (Adults_t/2) \times Per\ capita\ birth\ rate \times Density\text{-}dependent\ negative\ feedback_t \tag{9.59}$$

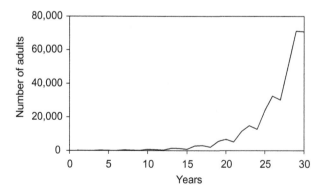

FIG. 9.25

Amphibian species survival model. Simulation results of the second intermediate developmental model (see Fig. 9.22B for a sketch of the associated conceptual model).

Since we now have quantified *density-dependent negative feedback*, we would expect our third, and last, intermediate developmental model to generate essentially logistic growth, with fluctuations about this general trend since we did not initialize the population in the stable stage distribution. Observing simulated population growth, we note the model generates this expected pattern (Fig. 9.26). We also note that the simulated population stabilizes at about 172 adults, which seems reasonable, given that historical estimates of the number of adults have fluctuated between 80 and 280 (see background information).

Having completed the quantification of our model, prior to applying the model to address the problem at hand, we would like to get a feel for how the uncertainty associated with the estimate of juvenile survival (mentioned in the background information) might influence our simulation results. To investigate this, we will conduct a sensitivity analysis consisting of six simulations in which we increase the value of the annual survival rate of juveniles from 0.5% (*juvenile mortality rate* = 0.995) to 3% (*juvenile mortality rate* = 0.97) in increments of 0.5%.

As the value of *juvenile mortality rate* was decreased from 0.985 to 0.97, the simulated number of adults stabilized at increasingly higher levels (between 87 and 193), whereas with higher values of *juvenile mortality rate*, the simulated number of adults declined continuously (Fig. 9.27). Given the sensitivity of population dynamics to juvenile survival, and given the year-to-year variability in juvenile survival, it seems reasonable to assume that the population may increase during some years and decrease during other years. The likelihood of species survival will depend on the frequency of occurrence of good and bad years. We will want to take this uncertainty into consideration when we apply our final model to address our modeling objective.

Final model: To use our final model to address the objective of the modeling project, that is, to estimate the likelihood of species survival in the face of the uncertainty associated with juvenile survival, we will use a stochastic version of our model. Specifically, we will represent *juvenile mortality rate* as a stochastic variable rather than as a constant. Each year during a simulation, we will draw the value of

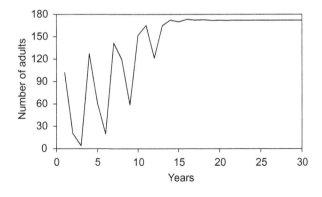

FIG. 9.26

Amphibian species survival model. Simulation results of the third, and last, intermediate developmental model (see Fig. 9.22C for a sketch of the associated conceptual model).

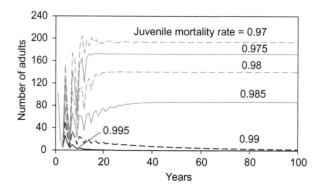

FIG. 9.27

Amphibian species survival model. Results of sensitivity analysis comparing population dynamics assuming *juvenile mortality rate* is equal to 0.97, 0.975, 0.98, 0.985, 0.99, and 0.995.

juvenile mortality rate from a uniform distribution on the interval from 0.97 to 0.995. Our choice of a uniform distribution, rather than, for example, a normal distribution, is arbitrary. If we had sufficient field data, we could try to fit some statistical distribution to the data, or we could draw a random variate from a cumulative frequency distribution based on the data (see Chapter 5, Section 5.5.1). Our experimental design will consist of a series of 100, 100-year, replicate stochastic (Monte Carlo) simulations, which we will assume is an appropriate number of replicate stochastic simulations (see Chapter 5, Section 5.6.2). We will calculate the likelihood of species survival as the proportion of the simulations during which number of adults never becomes less than 30 individuals. Our choice of 30 adults as the threshold for species survival is arbitrary, but we will assume that it is appropriate within the context of our problem.

  Our simulation results suggest that the likelihood of species survival is zero. Furthermore, all of the simulated populations decreased from the initial 102 adults to about 20 adults after 1 year to about four adults after 2 years. We suspect that these results may indicate a "start-up" problem. The behavior of any model during the first several time steps may characterize a "transition period" in which the state of the modeled system transitions from its initial state, which we defined, to a state or series of states that result from the dynamic interaction of the model components. That is, model behavior during this transition period might be an artifact of the particular way in which we initialized the model (in our case, with 102 adults but no larvae or juveniles). To confirm our suspicion of a possible "start-up" problem, we will run an additional set of 100 replicate stochastic simulations in which we will wait until year 25 to record simulation results. We have chosen to wait until year 25 because we notice that, in the deterministic version of our model, the proportions of individuals in each stage class no longer are changing by the time we reach year 25 (Fig. 9.28). Results of these additional simulations indicate that the likelihood of species survival has increased markedly to 77%. Thus, our suspicion of a "start-up" problem is confirmed.

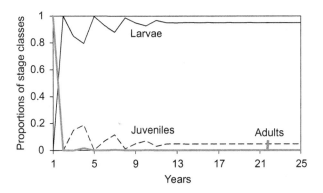

FIG. 9.28

Amphibian species survival model. Simulated proportions of individuals in each stage class (larvae, *black line*; juveniles, *dashed line*; adults, *gray line*) using the deterministic version of the model.

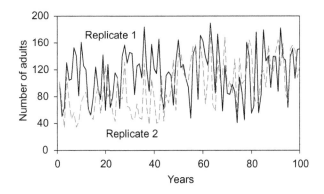

FIG. 9.29

Amphibian species survival model. Temporal trends in the number of adults from each of two typical replicated stochastic simulations. In these simulations, the amphibian population survived (the number of adults never fell below 30).

To avoid this "start-up" problem, we will initialize our model with individuals distributed according to the stable stage-class distribution, which we determined using the deterministic version of our model (0.9509 larvae, 0.0476 juveniles, 0.0015 adults; Fig. 9.28). Thus, in addition to the 102 adults reported in the field data, we will assume that there also are 3267 juveniles and 65,338 larvae in the population. We will repeat our experimental design (a series of 100, 100-year, replicate stochastic simulations) with these new initial conditions, and base our final estimate of the likelihood of species survival on the results of these simulations, which suggest that the likelihood of species survival is 72%. Temporal trends in the number of adults from each of two typical replicated stochastic simulations are shown in Fig. 9.29.

As with our previous two examples, these simulation results from our initial experimental design likely raise some interesting new questions, which we leave to your imagination and ingenuity to identify and explore. We also leave you, as before, with task of formally communicating the results of this modeling project.

Before concluding this chapter, we hasten to add a point worthy of note regarding the simulation "start-up" problem. In some cases, the transient model behavior exhibited during the first portion of simulated time may not be interpreted as a problem at all. Indeed, the "transition period" may characterize the model behavior in which we are most interested. In such cases, we usually have good information about the current state of the system-of-interest, or some specific reason within the context of our problem for having specified particular initial conditions as a point of departure for observing simulated system behavior.

Reflections

How good ("valid") are models?

10

Chapter outline

Developments in Environmental Modeling, Volume 31, ISSN 0167-8892, https://doi.org/10.1016/B978-0-444-64163-2.00010-4

10.1 **Overview**

There remains considerable confusion in the scientific literature regarding the appropriate ways to evaluate how "good" a model is. The process of evaluating a model is without a doubt the most controversial step in the modeling process. This confusion arises more from semantics and differences in philosophical viewpoints than from debates over specific methodologies used. Regarding semantics, the term "validation" takes center stage, followed by terms such as "verification," "confirmation," "corroboration," "credibility," "qualification," and "evaludation" (not a typo, see Augusiak et al., 2014), which have been used in attempts to place "validation" within an appropriate context. Regarding philosophical viewpoints, some modelers have stated that "verification" and "validation" of numerical models of natural systems are impossible, but that their "confirmation" is possible (e.g., Oreskes et al., 1994). Oreskes et al. (1994) based their statements on an examination of the philosophical basis of these terms, thus their philosophical viewpoint is inextricably tied to their semantics. Other modelers have stated that "validation" simply means that a model is acceptable for its intended use because it meets specified performance requirements (e.g., Rykiel, 1996). Rykiel (1996) provided his statement while discussing the meaning of "validation" within the context of ecological simulation models. Still other modelers have coined the new term "evaludation" to avoid use of "validation" per se by combining it with the more neutral term "evaluation," the intent being to keep the idea of "validation" visible while softening its connotation (Augusiak et al., 2014).

As we stated in Chapter 6, we prefer to refer to the process of determining how "good" a model is simply as "model evaluation." For us, the evaluation process, via

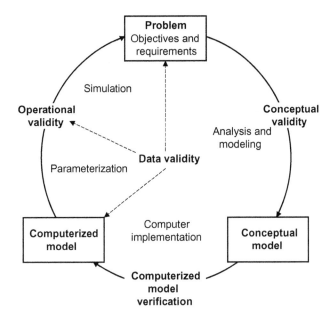

FIG. 10.1

A diagrammatic representation of the modeling cycle and the position of operational, conceptual, and data validation processes.

Modified from Sargent, R.G., 1984. A tutorial on verification and validation of simulation models. In: Proceedings of the 16th conference on Winter simulation. *IEEE Press, as represented by Rykiel (1996).*

the four steps described in Chapter 6, is focused on assessing usefulness of the model in meeting specific objectives of the modeling project. Thus, "good" is synonymous with "useful." Nonetheless, an appreciation of differences in semantics and philosophical viewpoints among modelers with regard to the process of model evaluation is a requisite for understanding the ecological modeling literature. In the following five sections of this chapter, we discuss these differences in (1) semantics and (2) philosophical viewpoints, present some (3) methodological approaches and (4) specific procedures associated with model evaluation, and consider some (5) evaluation criteria. We conclude with some final thoughts on the process of model evaluation.

We have drawn heavily on the classic papers by Oreskes et al. (1994) and Rykiel (1996) and the excellent recent review paper by Augusiak et al. (2014) for much of the following material. To help avoid getting lost in all of the jargon, we have redrawn the diagrammatic representations used by Rykiel (1996, Fig. 1, p. 234) and by Augusiak et al. (2014, Fig. 1, p. 121) to illustrate how their terminology and ideas on model evaluation fit into the general modeling cycle (Figs. 10.1 and 10.2, respectively). We will consider the details of these diagrams in Section 10.4 when we discuss methodological approaches. You will notice the general similarity of these two diagrams to each other and to the diagram we

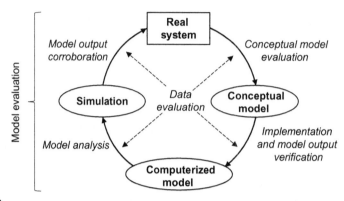

FIG. 10.2

A simplified representation of the modeling cycle, consisting of the four main steps of model development and their corresponding elements of evaludation. The terms in italics comprise the terminology suggested to organize and communicate model evaludation. The four elements of model development were, in the context of model quality assurance, suggested by Refsgaard and Henriksen (2004) and Schlesinger (1979). Their definition of quality assurance corresponds to what here is referred to as "evaludation" and what so far in ecological modeling usually has been implicitly referred to as evaluation and/or validation.

Modified from Augusiak, J., Van den Brink, P.J., et al., 2014. Merging validation and evaluation of ecological models to 'evaludation': a review of terminology and a practical approach. Ecol. Model. 280, 117–128.

presented at the beginning of our chapter on model evaluation (Fig. 6.1). We suggest that you refer back to these three figures often as you are reading the material in the rest of this chapter.

10.2 **Semantics**
10.2.1 **Verification**

In most dictionaries, "verify" (from Latin, *verus*, meaning true) means to ascertain or prove the truth. Based on this definition, to say that a model has been verified means that is has been shown to be true, that its truth has been demonstrated. The unfortunate connotation within the context of ecological modeling is that this implies the absolute reliability of the model as a basis for decision-making. But it is impossible to demonstrate the truth of any proposition, except in a closed system. This conclusion derives directly from the laws of symbolic logic. Given a proposition of the form "p" necessitates "q," we know that if "p" is true, then "q" is true if and only if the system that this formalism represents is closed. Numerical models may contain closed mathematical equations that are verifiable, and algorithms within a computer program are verifiable. Mathematical equations are subject to verification because they are part of closed systems that include claims that are always true as a function of the meanings assigned to the specific symbols used to express them. "Two plus two equals four" is a verifiably true statement. However, models that use these ver-ifiable mathematical equations never are closed systems. They require input param-eters that are known incompletely because observations and measurements of variables invariably are based on inferences and assumptions. In this respect, data might be viewed as inference-laden representations of the natural phenomena to which we have incomplete access. Many inferences and assumptions can be justified based on personal experience, and the associated uncertainties sometimes can be es-timated, but the degree to which our assumptions hold in any new study can never be established a priori. Thus, incomplete knowledge and the resulting assumptions we make render all natural systems open.

Within the context of simulation modeling, we might more usefully say that "verification is a demonstration that the modeling formalism is correct" (Fishman and Kiviat, 1968, cited in Rykiel, 1996). Based on this definition, we might identify two types of verification errors: mechanical and logical. The first involves debugging a computer program and showing that the mathematics is mechanically correct. A more subtle and more difficult verification problem is showing that the program logic is correct. Some program errors only appear under particular circumstances that may not occur often, and might not have been anticipated. Verification is a technical matter that relates to how faithfully and accurately ideas are translated into mathe-matical equations and computer code. For large models, it is extremely difficult to verify that the model is entirely accurate and error free under all circumstances, and that modifications of the existing code have only the intended effect. Thus, models generally are verified for the normal circumstances in which they are expected to be applied, and such verification is presumed inapplicable if the model is run outside of these circumstances. It is important to distinguish between verification of the logic that relates to computer program operation from verification of the logic that refers to the ecological logic used in structuring the model. Verification of computer model is quite different from showing that the computer model is a useful representation of the real system. Also worthy of note, is that the congruence between a numerical solution produced by a model and an analytical solution entails nothing about the correspon-dence of either one to material reality.

10.2.2 Calibration

Calibration simply refers to the process of adjusting values of model parameters to improve agreement between model output and observations from the real system. Calibration should be distinguished from the process of determining initial values of model parameters, which is parameter estimation. In essence, model calibration is the process of making model output as consistent as possible with data sets from which the model parameters initially were estimated. Thus, we could say that a calibrated model is empirically adequate. However, as more data become available, calibrated models invariably need additional refinements, suggesting that the empir-ical adequacy of numerical models necessarily is "forced" (Bas van Fraassen, 1980, cited in Oreskes et al., 1994). This fact has serious implications for use of any calibrated model for predictive purposes. (Consider stating that your model has "forced empirical adequacy.") Even if model results are consistent with present and past observational data, there is no guarantee that the model will perform equally well when used to predict the future. For example, there may be small errors in input data that do not impact the fit of the model under the time frame for which historical data are available, but which, when extrapolated over larger time frames, do generate significant deviations.

10.2.3 Validation

In most dictionaries, "valid" means to have a sound basis in logic or fact. In contrast to the term "verification," the term "validation" does not necessarily denote an establishment of the truth (although truth is not precluded). Rather, it denotes the establishment of legitimacy. A valid argument is one that does not contain obvious errors of logic. By analogy, a model that does not contain known or detectable flaws and is internally consistent can be said to be valid. Thus, the dictionary definition of valid might be applicable for assertions about a generic computer code, but clearly is misleading if used to refer to the actual model results in any particular situation. Model results may or may not be valid, depending on the quantity and accuracy of input parameters. The most common method of validation involves a comparison of numerical model output to data from the field or laboratory. But agreement between model output and field or laboratory data in no way demonstrates that the model is an accurate representation of the real system. Validation within this context simply represents consistency between or among data sets. Such consistency entails nothing about the reliability of the model in representing natural phenomena. The idea that a model accurately represents "actual processes occurring in a real system" is not even a theoretical possibility.

Within the context of simulation modeling, as with verification, validation is better understood as a process that results in an explicit statement about model behavior. We might more usefully say that "validation is a demonstration that a model within its domain of applicability possesses a satisfactory range of accuracy consistent with the intended application of the model" (Sargent, 1984; Curry et al., 1989, cited in Rykiel, 1996). This demonstration indicates that the model is acceptable for its intended use, not that it embodies any absolute truth, nor even that it is the best model available. Validation demonstrates that a model meets some specified performance standards under some specified conditions. It is often overlooked that the "specified conditions" include all implicit and explicit assumptions about the real system the model represents, as well as the environmental context of the model. That is, a model is declared validated within a specific context which is an integral part of certification. If the context within which the model is used changes, the model must be revalidated. However, that does not invalidate the model for use within the context in which it was originally validated. Validation is a yes or no proposition in the sense that a model does or does not meet specified validation criteria. These criteria may include requirements for statistical properties (e.g., goodness-of-fit) of data generated by the model, and thus are not necessarily deterministic. However, ambiguous situations can arise when the model meets some but not all criteria. Criteria may need to be prioritized, and the model may be validated with respect to these priorities. Because modeling is an iterative process, validation criteria may evolve along with the model.

10.2.4 Confirmation

Philosophers have developed the theory of confirmation to deal with the logical fallacy of "affirming the consequent." Recall that we mentioned in the previous section that the most common method of validation involves a comparison of numerical model output to data from the field or laboratory. But we concluded that agreement between model output and field or laboratory data in no way demonstrates that the model is an accurate representation of the real system. This conclusion is derived directly from the logical fallacy of affirming the consequent. As an example of affirming the consequent, consider the following situation. I tell you that I will stay home tomorrow if it is raining. Tomorrow you find that I am at home, therefore, you conclude that it is raining. Clearly, this is faulty logic. I might have decided to stay home even though it is not raining. To claim that a model is a correct representation of the real system because model output matches empirical data is to commit the fallacy of affirming the consequent. If a model fails to reproduce observed data, then we know that the model is faulty in some way, but the reverse is never the case. Thus, it has been suggested that we should refer to such a match between model output and empirical data as confirmation (Oreskes et al., 1994). The greater the number and diversity of confirming observations, the more probable it is that the model is not flawed.

10.3 Philosophical viewpoints

Ecological modelers have been discussing model validation since the1960s. That is, essentially since the "birth" of ecological modeling as a recognized scientific discipline. Development of the ideas and methodological concepts for validating ecological models has shifted direction several times since then. Some have argued that one of the main reasons for disagreements concerning semantics and methodological approaches lies in the philosophical views on how science is performed and, in turn, what validation means in science in general (for an excellent recent review, see Augusiak et al., 2014).

Logical empiricism dominated scientific conduct from the middle of the 19th century to the middle of the 20th century. This school of thought favored inductive inferences building from single observations and/or experiments to universal statements, such as hypotheses or theories (Barlas and Carpenter, 1990; Refsgaard and Henriksen, 2004). Proposed hypotheses or theories then are tested in experiments that are designed to confirm or refute the general statement at hand. From a model validation perspective, such an approach would make the process of validation formal and algorithmic. Under such premises, models would be assumed to be objective and absolute representations of the modeled system, such that they only could be either true or false. It seems that many nonmodelers view validation from this perspective.

Critics of this approach argue that theories only can be falsified and never verified (Kuhn, 1962; Popper, 1959). They promote a more deductive approach toward science in which inferences are drawn from universal statements, such as theories or hypotheses, to make more specified statements about the matter at hand. Conclusions are derived logically from several statements, and predictions of empirical patterns are formulated as deductive consequences from theories or hypotheses. If those conclusions and predictions can be shown to be true, the overarching hypothesis is deemed to be corroborated or confirmed (Popper, 1959). The larger the number of confirming observations, the more credible the hypothesis is deemed to be. However, no matter how many times the hypothesis is confirmed, there is always a chance that an observation can be explained by more than one hypothesis. Furthermore, a single falsifying observation is sufficient to reject the hypothesis.

From this deductive perspective, validation becomes a less formal process because a valid model is assumed to be only one of several probable representations of a real-world process. One model may be preferred over alternative models for a particular use, but no model can claim absolute objectivity because each embodies the subjectivity of the modeler or modelers who constructed it (Barlas and Carpenter, 1990; Oreskes and Belitz, 2001; Oreskes et al., 1994). Models are neither true nor false but lie on a continuum of usefulness for which credibility can be built up only gradually (Barlas and Carpenter, 1990; Rykiel, 1996). Thus, assessment of the validity of a model is changed from a true or false question to an assessment of how credible the model is in view of existing evidence.

However, falsifiability is not accepted as the ultimate, or even sufficient, criterion for scientific testing by all philosophers of science (Thagard, 1988, cited in Rykiel, 1996). Neither is Popper's rejection of induction shared by all philosophers of science (e.g., Trusted, 1979, cited in Rykiel, 1996) and certainly not by all ecologists (e.g., Mentis, 1988, cited in Rykiel, 1996). These writers would claim that a single-minded focus on falsification results in a superficial treatment of a complex subject that includes, for example, much thought about what constitutes scientific explanation and what constitutes a theory (e.g., Suppe, 1989). Theories often contain unobservable components and thus cannot always be falsified directly (Mentis, 1988). For example, the competition coefficients in Lotka-Volterra population models are not observable quantities. Practically, all ecological simulation models contain unobservable quantities, quantities that only can be observed with significant error, and calculations that cannot be compared to data because no data exist. The impulse to falsify can result in "naive falsification" (Rykiel, 1996). For example, the fact that the carbon budget of the earth does not balance according to current data could be taken naively as a falsification of the ecosystem theory that states that the earth is an ecologically closed system with respect to matter.

An overemphasis on the philosophical belief that scientific hypotheses cannot be proven but only disproved may contribute to the misunderstanding of model validation in another way. In one sense, the falsification argument fails to acknowledge the physical reality of the world. A premise can be proven to be true by showing that whatever is stated in the premise actually can be done or is otherwise objectively verifiable, even when scientific understanding is incomplete or even lacking (Rykiel, 1996). Such a demonstration is commonly termed "proof of concept" in the engineering jargon. This idea really is the basis for validation testing via the comparison of simulated data versus observed data.

The scientific method, hypothesis formation, analysis, and practice, cannot be reduced to a series of unequivocal steps which anyone can follow to make new discoveries, enunciate new theories, and validate models. Model evaluation brings to bear all of the subjective and objective elements of conscious and unconscious thought and reasoning of which the individual scientist is capable. In the words of Goodman (quoted in Thagard, 1988, cited in Rykiel, 1996): "A rule is amended if it yields an inference we are unwilling to accept; an inference is rejected if it violates a rule we are unwilling to amend. The process of justification is the delicate one of making mutual adjustments between rules and accepted inferences; and in the agreement achieved lies the only justification needed for either."

10.4 **Methodological approaches**

Given the variety of opinions regarding what constitutes a "valid" model, it is not surprising that there is a variety of proposed methodological approaches to assess how "good" a model is. However, as we mentioned at the beginning of our chapter on model evaluation (Chapter 6), in spite of debates over the semantic and philosophical details concerning model validation, we believe the basic ideas involved in evaluating the usefulness of a model are easy to understand. To support our belief, we first summarize the ideas embodied in the diagrammatic representations of the modeling cycle presented by Rykiel (1996) and by Augusiak et al. (2014), which we introduced briefly at the beginning of this chapter (Figs. 10.1 and 10.2, respectively). We believe that these authors provide an excellent summary of the ideas on model validation from an ecological perspective, and also demonstrate how the basic ideas involved in the evaluation of ecological models have not changed drastically over the past two decades. We then take a moment to call attention to the similarity of the ideas embodied in these two approaches and in the approach that we presented in our chapter on model evaluation (Fig. 6.1).

10.4.1 **The modeling cycle of Rykiel**

Rykiel (1996), while acknowledging that the validation process can encompass a large number of tests, suggested that for ecological models it seemed sufficient to consider a minimal partitioning of validation activities. Following Sargent (1984),

Rykiel described a simplified version of the modeling process that identified three principal areas where validation may be needed: (1) operational validation, (2) conceptual validation, and (3) data validation. In all of these cases, specific criteria for validation must be specified. Validation does not require that the model apply to more than one case unless that criterion is asserted as a validation requirement. That is, generality is no more required of a simulation model than it is of a regression equation. The testing procedures are used during model construction, at model completion, and for model qualification.

Operational validation

Operational validation is a test protocol to demonstrate that model output meets performance standards required for the purpose of the model. This is the purpose of prediction of Caswell (1976) and includes the purpose of projection (the qualitative correspondence with the event dynamics) of Rykiel (1984). This process is a pragmatic approach to validation because it is concerned primarily with how well the model mimics the system, regardless of the mechanisms built into the model. Statistical tests of comparisons between simulated and real data are used widely to evaluate model behavior (Mayer and Butler, 1993; Power, 1993). If model output corresponds with observed data, then the model is an adequate representation of the system. However, such a correspondence does not guarantee that the scientific basis of the model and its internal structure correspond to actual processes or cause-effect relationships that are operating in the real system. That is, such tests cannot demonstrate logical validity of the scientific content of the model (Oreskes et al., 1994). Although it may seem paradoxical, in general, it is not true that good predictions can only be obtained from a model that is mechanistically correct. Operational validation is the engineering side of the validation process that seeks to demonstrate model capability and continues during the model-building process until a prescribed level of performance is attained. Failure to meet an operational validation test also can reveal underlying conceptual problems (e.g., Fleming and Shoemaker, 1992).

Conceptual validation

Conceptual validity means that the theories and assumptions underlying the conceptual model are correct, or at least justifiable, and that the model representation of the problem or the system, its structure, logic, mathematical, and causal relationships, are reasonable for the intended use of the model. Conceptual validity depends on providing a scientifically acceptable explanation of cause-effect relationships that are included in the model. Alternatively, justification is given for using simplifications of known processes, and for conjectured relationships for poorly known processes and mechanisms. Such justification may extend to providing a rationale for leaving out processes known to be involved in system behavior, and for using representations known to be ecologically false, that is, for using a conceptually invalid model. Because the model is an abstraction of the real system, many components and processes must be left out for the purposes of a particular model. Conceptual validity, considered as acceptable justification of the scientific content of the model, does not guarantee that the model will make accurate predictions.

Data validation

Data are not an infallible standard for judging model performance (Fagerström, 1987). Rather, the model and the data are two moving targets that we try to overlay one upon the other. Validation of the data certifies that the data meet a specified standard (quality assurance/quality control). We cannot assume that the data accurately represent the real system and therefore constitute the best test of the model. Not only validity of the data, but also validity of the interpretation of the data, must be demonstrated. Holling (1978) cited an instance where a model was reanalyzed for two months after which it finally was determined that the data had been misinterpreted and the model was not at fault. The relative inaccuracy and imprecision of ecological data also places limits on model testability. Obviously, computer simulation models cannot be expected to provide results that are more accurate and precise than the data that are available. Conversely, it can be argued that the model may be a better representation of reality than are the data, which are limited by our technological abilities for measurement and are subjectively biased by our perceptions of the system.

10.4.2 The modeling cycle of Augusiak et al.

Augusiak et al. (2014) followed pragmatic recommendations from various scientific fields to split the model evaluation process into several steps (Barlas, 1996; Refsgaard and Henriksen, 2004; Rykiel, 1996; Sargent, 2013). However, to describe the process, they combined the term "validation" with the term "evaluation" to introduce the new term "evaludation." They defined "evaludation" as "the entire process of establishing model quality and credibility throughout all stages of model development, analysis, and application." They coined this term to (1) avoid using "validation" itself while maintaining its visibility, (2) link the request for validity assurance with the more neutral, multistep process of evaluating the quality of not only model output but also all other relevant aspects of the modeling process that can affect model credibility and validity, and (3) emphasize the multicriteria character of model assessment. Evaluation consists of six elements: (1) data evaluation, (2) conceptual model evaluation, (3) implementation verification, (4) model output verification, (5) model analysis, and (6) model output corroboration. The authors suggest that all of these elements are necessary to communicate which uncertainties have to be faced at different stages of the modeling cycle, which evaluation tools are applied for these elements, and which measures should or could be taken to reduce uncertainties.

Data evaluation

Data evaluation is defined as the critical assessment of the quality of the numerical and qualitative data that are used to parameterize the model, both directly and inversely via calibration, and of the observed patterns that were used to design overall model structure. Thus, data refer not only to numerical data, but also qualitative data such as expert knowledge for which no hard numbers exist. Computer models can take into account such qualitative expert knowledge in the form of probabilistic if-then rules. The term data also refers to patterns (Grimm and Railsback, 2012), which are general trends and signals in data, observations, and empirical knowledge. The appropriateness, accuracy, and availability of data used throughout model development are major sources of uncertainty and often are the reason for failed attempts to validate a model (Sargent, 2013). Quality of available numerical data can be corrupted by measurement errors, by flawed experimental designs, and by the natural heterogeneity and stochastic variability inherent in environmental systems (Gass, 1983; Wang and Luttik, 2012). Likewise, expert knowledge and the detection of patterns are notoriously prone to bias and therefore must be treated with particular caution. Another aspect to watch carefully concerns extrapolation of data from one situation to another. This can include use of laboratory data to estimate analogous effects in the field, use of data from another climate zone or from another species, and use of experimental data that are collected only during a particular period or in a particular area (Fagerström, 1987; Topping et al., 2012).

Conceptual model evaluation

Conceptual model evaluation is defined as the critical assessment of the simplifying assumptions underlying the design of a model. Evaluating the conceptual model means to explicitly list, discuss, and justify spatial and temporal scales, choice of entities and processes, stochasticity and heterogeneity, local versus global interactions, and representation of environmental drivers. Conceptual model evaluation also includes assessment of whether the structure, underlying theories, concepts, assumptions, and causal relationships are reasonable to form a logically consistent model. Limitations arising from simplifying assumptions need to be considered and justified. Conceptual model validity is affected primarily by the subjectivity of the modeler, his/her incomplete knowledge of the problem at hand, and quality of available data. As a guiding principle, Occam's razor should be applied to ensure that the chosen model complexity does not introduce avoidable uncertainty (Beck et al., 1997; Clark, 2005; Jakeman et al., 2006).

Implementation verification

Implementation verification is defined as the critical assessment of (1) whether the computer code for implementing the model has been thoroughly tested for programming errors and (2) whether the implemented model performs as indicated by the model description. This element of evaluation is concerned primarily with checking the computer code for errors and oversights. However, even an error-free program code might not implement the model as intended or described, which could be due to ambiguities in the model description or misinterpretations of ready-to-use procedures in the software platform used. Although implementation verification focuses primarily on technical aspects of a modeling exercise, it also is essential for assessing whether a model is of sufficient realism and quality. Model output might look promising, but without a thorough evaluation of the implementation procedure the model might lead us to erroneous conclusions because the model is not working as we think it does.

Model output verification

Model output verification is defined as the critical assessment of (1) how well model output matches observations from the real system and (2) to what degree calibration and effects of environmental drivers are involved in obtaining good fits of model output and data. When comparing model output to data from the real system, initial conditions of the model, simulated environmental conditions, and protocols for sampling simulated data should match those underlying data from the real system as closely as possible (Zurell et al., 2012). For complex models, this can be a difficult task (Topping et al., 2012). In addition to comparing simulated data with real data per se, it is important to consider how much calibration was involved in achieving sufficient similarity between the two. The more parameters that are calibrated, the higher the risk that successful output verification is achieved via an unrealistic combination of parameter values. Likewise, a good match of model output and observations from the real system might be imposed via representation of strong environmental drivers (e.g., the weather conditions), in which case model output is primarily the result of dynamics of external drivers rather than resulting from a realistic representation of system structure. What is considered a sufficiently similar match between model output and observations from the real system to warrant output verification depends on model objectives and the current level of understanding of the system in question.

Model analysis

Model analysis is defined as the assessment of (1) how sensitive model output is to changes in model parameters (sensitivity analysis) and (2) how well the emergence of model output has been understood. Testing model sensitivity is essential since a good match of model output and data might result from the calibration of several model parameters. The match might vanish as soon as one or more or the calibrated parameters are changed. Sensitivity analysis identifies subsets of parameters that have strong effects on model outputs, thereby indicating which processes are most important for further consideration. Insights into model behavior provided by sensitivity analysis help prevent viewing the model as a "black box" by helping to explain how model output emerged. This understanding of the sensitivity of model output also provides information that often is important for supporting model-based management decisions, since it indicates the uncertainty associated with model predictions.

Model output corroboration

Model output corroboration is defined as the comparison of model predictions with independent data and patterns that were not used, and preferably not even known, while the model was being developed, parameterized, and verified. The emphasis on new, independent data is important. A model often will reproduce data used during model development. But comparisons of model output to data used in model development entail the unavoidable risk that the model might produce the right output for the wrong reasons. One form of new data consists of results from new and specifically designed field experiments. This is one of the most common interpretations of validation: a model is considered "valid" if it made predictions that were confirmed by subsequent field experiments. However, such field experiments usually are not feasible for most ecological systems. Thus, for model output corroboration we usually have to resort to comparing model predictions with data that already exist, but that have not been known or used during model development. It is important to maintain the distinction between data used for model verification and data used for corroboration of model output. Data used for verification of model output can be data already used in model development. Data used for corroboration of model output must be data not already used in model development. Note that a model still can be considered sufficiently realistic and appropriate to fulfill its purpose even if corroboration was not possible due to lack of resources or data.

10.4.3 Similarities in modeling cycle representations

We now compare the perspectives embodied in Figs. 10.1, 10.2, and 6.1 regarding how the process of model evaluation fits into the general modeling cycle. You will notice the general similarity of these three diagrams. All of them begin with questions based on observations from the real system, which lead to development of a conceptual model of the system-of-interest, from which a quantitative computer model is developed, which then is evaluated prior to its application to address the original questions. We provide a point-by-point comparison of these schemes below.

Conceptual validity (Fig. 10.1) means that the theories and assumptions underlying the conceptual model are correct, or at least justifiable, and conceptual model evaluation (Fig. 10.2) is defined as the critical assessment of simplifying assumptions underlying the design of a model. We have noted that the six steps in conceptual model formulation (Fig. 4.1) are the most intellectually challenging in the modeling process, and that the best basis for the often highly subjective decisions regarding choice of model components is a thorough familiarity with the real system, although prior modeling experience also is an asset (Chapter 4).

Computerized model verification (Fig. 10.1) involves debugging a computer program and showing that the mathematics is mechanically correct, as well as showing that the program logic is correct. Implementation verification (Fig. 10.2) is defined as the critical assessment of (1) whether the computer code has been thoroughly tested for programming errors and (2) whether the implemented model performs as indicated by the model description. We have noted that after parameterization of the model, but prior to formal running of the baseline simulation, we must verify that the computer code is correct ("debug" the code). That is, we must make sure that the computer program is doing what we want it to do (e.g., if we want the code to add $2+2$, the code must generate a result of 4) (Chapter 5).

Operational validation (Fig. 10.1) is a test protocol to demonstrate that model output meets performance standards required for the purpose of the model. Model analysis (Fig. 10.2) is defined as the assessment of (1) how sensitive model output is to changes in model parameters (sensitivity analysis) and (2) how well emergence of model output has been understood, whereas model output corroboration (Fig. 10.2) is defined as the comparison of model predictions with independent data that were not used while the model was being developed, parameterized, and verified. We have noted that the goal of model evaluation is to assess relative usefulness of the model for a specific purpose (Chapter 6). We also noted that model evaluation involves a broad array of qualitative as well as quantitative aspects of model structure and behavior, including sensitivity analysis (Fig. 6.1).

10.5 **Procedures**

Procedures for testing validity of ecological simulation models include both qualitative and quantitative measures of model performance, and project-specific tests are common. Although commercial and government systems modeling projects often require independent verification and validation of the resulting models (Lewis, 1992), there is no structured reporting requirement for verification and validation of ecological models. This is due in part to the fact that there is no institutional or scientific infrastructure for independent verification and validation of ecological simulation models, and due in part to the fact that ecological simulation models often are developed in an ad hoc fashion rather than as highly structured software development projects. Nonetheless, Rykiel (1996) has described a variety of validation procedures applicable to ecological models. He also has suggested a general scheme for deciding which types of procedures might be used based on amount of data available and level of understanding of the system being modeled (Rykiel, 1996, Fig. 2, p. 236) (Fig. 10.3).

10.5.1 **Face validity**

Knowledgeable people are asked if the model and its behavior seem reasonable "on the face of it." This test suggests whether model logic and input-output relationships appear reasonable given the purpose of the model. Some ecological models have high face validity by virtue of their longevity and wide spread use.

10.5.2 **Turing tests**

Knowledgeable individuals are asked if they can discriminate between model outputs and data from the real system. Requests of the evaluator to observe how closely simulated data and actual data match in a graphical display, for example, in a time

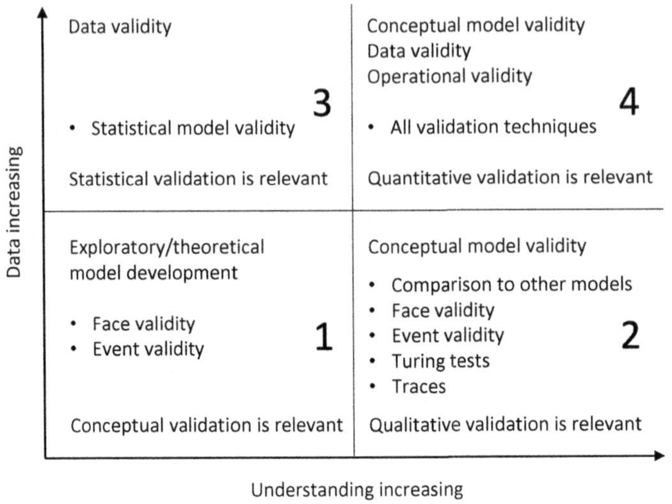

FIG. 10.3

Classes of modeling problems related to available data and understanding. Variation in amount of data available and level of understanding of the system influences the types of validation tests that can be conducted.

Modified from Holling, C.S., 1978. Adaptive Environmental Assessment and Management. John Wiley & Sons, Chichester, UK; Starfield, A.M., Bleloch, A.L., 1986. Building Models for Conservation and Wildlife Management. Macmillan.

series graph, are essentially unsupervised Turing tests combined with visualization. The subjective elements of this test need to be considered carefully (Mayer and Butler, 1993).

10.5.3 Visualization technique

Time series plots, state space phase plots, and other visual displays form the basis for comparisons between model outputs and data from the real system. Most often, validation is determined subjectively by a statement that summarizes the visual goodness of fit.

10.5.4 Comparison to other models

Output of one model can be compared to output of another model. In some cases, such as global climate models, this may be the principal means of evaluation (Cess et al., 1990).

10.5.5 Internal validity

A test data set consisting of initial conditions, parameter values, and input data for driving variables can be shown to produce a consistent output each time the model is run. This test is particularly applicable to stochastic models.

10.5.6 **Event validity**

A comparison between model outputs and data from the real system is made with regard to occurrence, timing, and magnitude of simulated and actual events. Event validity also may be interpreted as a qualitative validation in which the model is tested for its ability to reproduce proper relationships among model variables and their dynamic behavior rather than to accurately reproduce their quantitative values.

10.5.7 **Historical data validation**

When historical data exist, part of the data is used to build the model and part of the data is used to test if the model behaves as the real system behaves. This procedure also is referred to as data-splitting (Power, 1993).

10.5.8 **Extreme-condition tests**

The model should be capable of producing plausible output under extreme conditions or under combinations of conditions that are unlikely to occur in the real system. This test reveals if model behavior outside of the normal operating conditions is bounded in a reasonable manner.

10.5.9 **Traces**

The behavior of specific variables is traced through the model and through simulations to determine if the behavior is correct and if the necessary accuracy is obtained.

10.5.10 **Sensitivity analysis**

The same relationships that occur in the real system should occur in the model. Those parameters that cause significant changes in model behavior should be estimated with the greatest accuracy. Frequently, there is disparity between parameters to which the real system is sensitive and parameters to which the model is sensitive, but this issue is seldom mentioned.

10.5.11 **Predictive validation**

The model is used to forecast behavior of the real system and comparisons are made to determine if model predictions are the same as behavior of the real system. Real-system data may come from data sets not used in model development or from future observations of the system. The strongest case is when model output is generated before the real-system data are collected. Evaluation of model predictions often is considered as a kind of hypothesis testing.

10.5.12 **Statistical validation**

Statistical validation includes a variety of tests performed during calibration and operation of the model. Three cases occur most often. (1) The model produces output that has the same statistical properties as observations obtained from the real system. (2) The error associated with critical output variables falls within specified limits or acceptable limits. (3) Several models are evaluated statistically to determine which best fits available data.

10.5.13 **Multistage validation**

Validation methods are applied to three critical stages in the model building process. (1) During model design, model assumptions are evaluated based on theory, observations, general knowledge, and intuition. (2) During model implementation, model assumptions are tested empirically if possible. (3) During model operation, model input-output relationships are compared to those of the real system. The three validation steps correspond roughly to conceptual, data, and operational components of the validation process (Fig. 10.1).

10.6 **Criteria**

It is common for modelers to claim that a model has been validated without any reference to specific validation criteria. Often, it is stated that the model has been validated by comparison of model output with empirical data. Such statements indicate some sort of operational validation, but are otherwise meaningless until standards of comparison are specified. Furthermore, operational validation implies nothing about mechanistic soundness of the model. That is, ability of the model to predict independent data does not necessarily mean that the model contains cause-effect relationships that appropriately represent cause-effect relationships in the real system. This may or may not be a problem depending on objectives of the modeling project. Thus, not surprisingly, there is no standard set of criteria that is suitable for validation of all ecological models. Rykiel (1996) has provided some relevant and provocative thoughts regarding the various contexts within which model validation criteria might be considered.

10.6.1 **Qualification versus invalidation**

When operational validation is the objective, the difference between model qualification and model invalidation becomes an important distinction. Model qualification is testing aimed at determining the domain over which the model is applicable. The model is expected to fail outside this domain (or context). Consequently, the model may be subjected to increasingly stringent tests to determine conditions under which it fails to be a satisfactory simulator of the real system. When a model fails a validation test, several options are available. (1) The model may be recalibrated to improve its fit to data by changing parameter values. (2) The model may be modified structurally and conceptually by revising assumptions and by changing mathematical or logical representation of processes included in the model. (3) The application of the model may be restricted to a smaller domain where it is able to pass the validation test or where the particular test is not important. (4) Finally, failure to pass a validation test may be considered to invalidate the model. Invalidation implies that the model cannot pass a comparison test that is deemed essential for its credibility, acceptance, or usefulness. Notice that the implication is not that scientific content of the model is erroneous, it may simply be insufficient. Scientific content is embodied in assumptions, hypotheses, and theories on which the model is based and on the real system that the model is intended to simulate. For complex simulation models, falsification has little or no traditional meaning because the scientific content is too extensive to be proven wrong en masse. Invalidation can only show that a model does not or cannot meet its validation criteria. Parts of the model may well be operationally and conceptually acceptable even though the integrated model is unable to pass particular validation tests. Conversely, the integrated model may be validated even though parts of the model are scientifically invalid.

10.6.2 Validation and scale

Hierarchy theory indicates that the interpretation of mechanisms in the modeled system varies with the spatial and temporal scales at which a system is observed and modeled. Validation criteria likewise can vary with the spatial and temporal scales. For example, the level of mechanistic detail that is considered valid and the allowable error tolerances for simulating real-system behavior are likely to be quite different for a model of global carbon budget versus a model of carbon balance of a leaf. Consequently, our notion of what is acceptable depends not only on model objectives but also on the ecological scale at which the model is framed. Changes in scale may permit aggregations that, at face value, are conceptually erroneous, but that produce operational results which are acceptable.

10.6.3 Model validation versus theoretical development

There are many parallels between development and testing of models and development and testing of theories. A variety of studies suggest that prematurely imposing rigorous testing requirements can result in rejection of correct, or at least useful, theories (Loehle, 1987). Just as development of ecological theory can be stifled by an overemphasis on testing of hypotheses (Fagerström, 1987; Mentis, 1988), modeling and the benefits to be gained from it also can be stifled by an overemphasis on validation of models.

Neither falsification nor validation are required for development of useful theories, nor are they required for development of useful computer models

(Fagerström, 1987; Mentis, 1988; Thagard, 1988). Nature of the testing undertaken depends on purpose of the model. Purposes such as exploration, initial development, gaining new perspectives, among others, do not require model validation. Both theories and models often start off with erroneous elements which are discarded as development and understanding improve, and a significant loss of creativity occurs when either theory or models are not allowed to mature. Some things that are believed to be true cannot be validated or falsified in any easy, obvious, or immediate way. Caswell (1988) has argued forcefully that theoretical models are useful without any attempt at validation, and even when they are refuted.

10.6.4 Engineering validation versus scientific validity

Performance testing is limited fundamentally to showing that the model passes the validation tests devised for it. Certification is completed when the model is shown to meet its required specifications. This usually has been understood as some measure of agreement between simulated data and observed data and the judgment of experts. Ecological modelers do not describe in an engineering sense the required specifications of the models they build. Because there are no generally agreed validation criteria for ecological models, the best that can be done at present is for modelers to state explicitly what the validation criteria are and then leave it to users to judge if the criteria are adequate. The most common criteria at present are the "see how well simulated data match observed data" test and the "the model did a reasonable job of simulating ..." test in which the reader is asked to agree subjectively that the match is adequate (combinations of face validity, Turing test, and visualization). While validation testing may result in accumulating evidence that the scientific content of a model is correct, such testing cannot logically prove that the mechanisms contained in the model are scientifically complete and correct.

10.6.5 Validation and policy

To the extent that the technical meanings of verification and validation may be misconstrued, modelers themselves should take the lead in asserting restrictions and limitations of their models, and should draw some important lessons from Oreskes et al. (1994). (1) Make it clear that the terms verification and validation are used in a technical sense. (2) Carefully specify the context of the model. (3) Use model acceptability (Bart, 1995) and performance indices rather than simple declarations of validity to describe results of model testing to general audiences. Policies and decisions must be made and models, however imperfect, are needed to assist in the process. We have sufficient knowledge to accomplish many useful tasks and build a variety of devices whose very existence demonstrates that pragmatic operational validation is possible despite our philosophical uncertainties.

10.7 **Some final thoughts**

So what are we to make of this variety of (1) semantics, (2) philosophical viewpoints, (3) methodological approaches, (4) specific procedures, and (5) evaluation criteria associated with deciding how "good" a model is? It certainly can be argued that disagreements over the process by which we evaluate models are rooted in philosophical disagreements concerning the way we do science. Likewise, it certainly can be argued that confusing terminology is one of the main obstacles in understanding what is involved in the process of model evaluation. And, most certainly, philosophical debates and debates over terminology are not without merit. Such debates are thought-provoking, and they focus our attention on important issues. Ironically, it is precisely the lack of consensus on important issues that focuses our attention most effectively on them. That is, consensus is not necessarily good!

However, we would suggest that there is one general theme that pervades the material presented in this chapter. Stated very simply, appropriate evaluation criteria for a model depend on specific objectives of the modeling project (Grant, 1986; Grant et al., 1997; Grant and Swannack, 2008). This has been pointed out repeatedly, if at times not so simply, in the literature (Hoover and Perry, 1989; Mankin et al., 1977; Mayer and Butler, 1993; Rykiel, 1984, 1996; Augusiak et al., 2014). We can gain a deeper understanding of the model evaluation process by following debates about modeling philosophies and semantics, as long as we do not become enslaved by them. We need to trust our common sense. Below, we humbly present our, hopefully commonsensical, synthesis of the material in this chapter. We claim no originality in the ideas presented, only, perhaps, in the form of their presentation.

10.7.1 **Prerequisites for model evaluation**

The most common problem with evaluation of ecological simulation models is that we do not state the criteria we are going to use to evaluate them. There are no universally accepted standards for specifying evaluation criteria, nor should there be. Thus, before we begin the evaluation process, we should explicitly state (1) purpose of the model, (2) performance criteria that are required of the model for it to be declared acceptable for use, and (3) context of the model, that is, the domain over which the model is intended to operate. The latter two items seldom are specified for ecological models. Context, in particular, usually is overlooked. But it is impossible to evaluate a model without stating explicitly performance criteria and context.

Purpose

We construct ecological models both for scientific research and for environmental and natural resource management. In any case, our statement of purpose must include a clear and concise statement of the question or problem we want the model to address. That is, we need to identify what system we are modeling and what we are trying to learn about that system. Our statement of purpose provides the framework for model development, indicates the standards for model evaluation, and suggests the context within which simulation results will be interpreted.

Performance criteria

A totally objective and universally accepted approach to determining specific model performance criteria is impossible to imagine. Since models are approximations of reality, we always can establish performance criteria in such a rigorous form that they are impossible for any model to meet. On the other hand, we always can establish performance criteria in such a trivial form that they are meaningless. We described a variety of potential measures of model performance, both qualitative and quantitative, in Section 10.5. However, ultimately, we need to rely on our commonsense to establish performance criteria that will help us to assess usefulness and credibility of the model for achieving its stated purpose. Commonsensically, usefulness refers to the amount of new knowledge we can gain from the model and credibility refers to the extent to which model users have sufficient confidence in the model to base scientific or management decisions on model output. Obviously, both usefulness and credibility of a model are qualitative judgments that cannot be quantified in any absolute sense. Both are related inextricably to purpose of the model, amount of knowledge available about the system the model represents, and consequences of any decisions based on model output.

Context

Context of a model embodies all the assumptions under which the model operates. Context refers especially to those assumptions that remain implicit. For example, a model whose purpose is to simulate dynamics of vegetation in an alpine meadow might not include volcanic eruptions within its context. The model might assume that the vegetation dynamics are occurring in a nominal "noneruptive" environment. Thus, the model might not be acceptable for simulating vegetation dynamics following a volcanic eruption. The model would need to be reevaluated to see if the context within which it met its performance criteria could be expanded to include posteruption vegetation dynamics. In general, as a model passes reevaluation tests, the domain of applicability of the model increases. When the model fails a reevaluation test, its domain of applicability is restricted to the domain in which it has met all of its performance criteria. Obviously, it is impossible to define precisely the complete context of a model, and what assumptions we might leave implicit depends to a large extent on the audience to whom we are describing the model. Generally, the more multidisciplinary the audience, the larger will be the number of the assumptions that we should make explicit.

10.7.2 Importance/interpretation of model evaluation

Obviously, having devoted two chapters of this book to the topic, we think that the process of model evaluation is extremely important. But, just as both usefulness and credibility of a model are related inextricably to purpose of the model, relative importance of the various steps in the evaluation process also depend directly on purpose of the model. In this regard, we should be careful not to confuse the fact that we sometimes de-emphasize, or completely omit, some of the steps in Figs. 10.1, 10.2, and 6.1 with the fact we must evaluate carefully all models.

For example, near the beginning of Chapter 1 (Section 1.2.4), we suggested that simulation models might be placed along a continuum from "correlative" to "explanatory." Recall that we defined correlative models as those developed primarily to describe and summarize a set of relationships, without regard for appropriate representation of processes or mechanisms that operate in the real system. The goal is prediction, not explanation. We defined explanatory models as those developed primarily to represent internal dynamics of the system-of-interest appropriately. The goal is explanation through representation of causal mechanisms underlying system behavior, not prediction. We also mentioned that the classification relative to this dichotomy is based more on intent of the modeler, that is, purpose of the model, than on structure of the model per se. Thus, for models viewed as correlative, we might not evaluate reasonableness of model structure and ecological interpretability of functional relationships within the model. Likewise, for models viewed as explanatory, we might not be concerned about correspondence between model predictions and data from the real system.

We also should be careful not to confuse discussions about what kinds of models can versus cannot be "validated" with the fact that all models can and should be evaluated rigorously. For example, Rykiel (1996) suggested that when the principal purpose of a model is to describe or to systematize knowledge or to develop theory, validation is unnecessary and irrelevant. Within the model evaluation framework that we presented in Chapter 6, we would interpret this statement to mean that for explanatory models we would skip step 3 in Fig. 6.1. That is, as we stated in the previous paragraph, for explanatory models we would not examine correspondence between model predictions and data from the real system. But we certainly would assess reasonableness of model structure and interpretability of functional relationships within the model (step 1, Fig. 6.1). We also would evaluate correspondence between model behavior and expected patterns of model behavior (step 2, Fig. 6.1), which we identified during the final step of conceptual model development (Fig. 4.1). These expected patterns embody our hypotheses about how functional relationships within the model interact to generate dynamics of the system-of-interest. Almost surely we also would conduct a sensitivity analysis of the model (step 4, Fig. 6.1). We would be interested in both the range of conditions over which similar patterns emerge and the manner in which simulated patterns depart from expected (the hypothesized) patterns.

Thus, evaluation of purely exploratory models is as rigorous and important as evaluation of models that are intended for practical applications. But not only are procedural details different, interpretation of results of the evaluation process is different. Whereas models intended for practical applications can be evaluated and deemed acceptable for pragmatic purposes, theoretical acceptability always is provisional. And, to the extent that an exploratory model is a scientific experiment and represents a theoretical development, it could be argued that model developers do not have sole responsibility for evaluating the model (Rykiel, 1996). Rather, testing and evaluation of the model are within the purview of the scientific community. Both theories and models may be proposed even when their authors have undertaken no experimentation, but only have suggested what observations could be made to falsify the hypotheses embodied in their theory or model. A theory or model may explain previous observations, but it also may suggest other things that should be looked at and that may not yet have been measured, in which case there may be little or no data to be explained. The fact that a theory or model may be falsified in the future does not negate its contribution to present development of the field.

What good is modeling?

11

Chapter outline

11.1 Overview

In the previous chapter, we offered some reflections on the process of determining "how good" a model is. We stated that for us a "good" model is synonymous with a "useful" model. That is, useful in addressing the stated objectives of the modeling project. We focused primarily on the processes in the modeling cycle involved in model evaluation. In this chapter, we offer some reflections on the usefulness of the modeling process itself and consider how ecological modeling fits into the broader context of problem solving. More specifically, we (1) compare systems modeling to other problem-solving methods, (2) discuss ecological modeling as a problem-solving process, (3) consider the expectations we, as modelers, should have of ecological models, and (4) reflect on the expectations we, as environmental decision-makers, should have of ecological modeling.

As we suggested at the beginning of Chapter 1, we believe that the usefulness of ecological simulation modeling often results as much from the process (conceptual model formulation, quantitative model specification, and model evaluation) as from the product (the final model and its applications). We agree with Walters (1986) that the primary value of modeling in fields like ecology and natural resource management is not to make precise predictions, but rather to develop clear caricatures of nature against which to test and to expand experience. This certainly is not to say that simulation results produced by ecological models are without value in applied ecology and natural resource management. Ecological simulation models are recognized widely as valuable tools to aid in addressing an increasing variety of questions in applied ecological fields. However, we must form our expectations of ecological models in view of the uncertainties associated with their development.

11.2 Systems modeling compared to other problem-solving methods

The systems approach certainly is not the only useful approach to problem solving. Throughout the history of mankind, trial and error has been by far the most widespread and useful method. Unfortunately, for some problems, the appropriate trials are too long and the possible errors are too costly. The scientific method of solving problems emphasizes more disciplined observation and perhaps manipulation of particularly interesting parts of the world. Scientists formally interpret these observations in a variety of qualitative (description and classification) and quantitative (mathematical and statistical analyses) ways, depending on the type of problem, or type of system, with which they are dealing. Scientists and other problem solvers use the systems approach to integrate relevant information gained from trial and error, expert opinion, and the scientific method, in a form that facilitates formal description of the structure and dynamics of complex systems.

A precise definition of "complex system" is neither possible nor necessary. In Chapter 1, we suggested that a good working definition for our purposes is simply that a complex system is one whose behavior is impossible to understand intuitively or via our brain's unaided reasoning process. This inability to understand results more from indirect effects caused by the interconnectedness of system components than from the sheer number of system components. In fact, to relate the systems modeling approach to other formal methods of problem solving in a very general way, we might characterize systems in terms of number of components and degree of interrelatedness among components (Fig. 11.1). Problems related to systems with relatively few, highly interrelated components ("organized simplicity") can be addressed in an analytical form mathematically. Physicists deal with mechanical systems in which they are interested in this manner. Newton's laws of motion are an example. Problems related to systems with relatively many, loosely related components ("unorganized complexity") can be addressed statistically. For example, movement of gas molecules inside a closed, expandable container can be viewed as random and interesting system relationships can be described in terms of average temperature, pressure, and volume. But problems related to systems with relatively

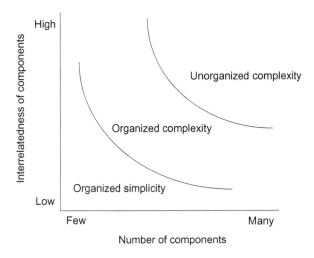

FIG. 11.1

General characterization of different types of systems in terms of number of components and degree of interrelatedness among components.

Modified from Weinberg, G. M. (1975). An introduction to general systems thinking (Vol. 304). New York: Wiley.

many, closely interrelated components ("organized complexity") cannot be addressed effectively by either of these two problem-solving methods. On the one hand, such systems usually cannot be solved mathematically because an analytical solution to the set of equations describing the system does not exist. On the other hand, dynamics of these systems cannot be represented statistically as average tendencies because the interrelatedness among components, or system structure, causes markedly nonrandom behavior. The systems modeling approach focuses specifically on these "intermediate" systems characterized by "organized complexity" in which system structure both controls and is changed by system dynamics.

From a slightly different perspective, we might compare methods of problem-solving in terms of relative level of understanding of the system-of-interest and relative amount of data available (Fig. 11.2). As used here, understanding refers to the informal integration of all that we have learned about the system by any means. Amount of data refers to the degree to which we have good data on all relevant aspects of the system. If we understand structure and general dynamics of the system and have good data on all important processes occurring within the system, we often can develop mathematical models and solve them analytically. If we have good data but lack understanding of the underlying system processes that generated them, we often can use statistical analyses to search for patterns in the data that will help us hypothesize the nature of these underlying processes. If we have relatively few data but at least some understanding of system structure and general dynamics, we often can use the systems modeling approach, consisting of systems analysis and systems simulation, to investigate our hypotheses about how the system works. Obviously, if

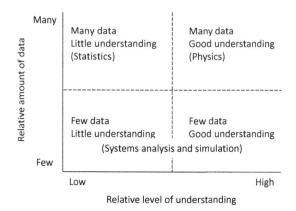

FIG. 11.2

Comparison of methods of problem-solving in terms of relative level of understanding and relative amount of data available on the system.

Modified from Holling, C.S., 1978. Adaptive Environmental Assessment and Management.
John Wiley & Sons, Chichester, UK; Starfield, A. M., & Bleloch, A. L. (1986). Building models for conservation
and wildlife management. Macmillan.

we cannot formulate useful hypotheses concerning system structure and function based on our current understanding, we should focus our efforts on further observation of the system.

Of course, in practice, domains of the various problem-solving methods overlap broadly. The most useful method for a given problem at a particular time depends on our conceptualization of the problem itself, which places us in one of the regions in Fig. 11.1, and on current state of knowledge about the problem within this conceptual framework, which places us in one of the regions of Fig. 11.2. Relative usefulness of the various methods also changes as we continue to work on a given problem, that is, as we increase our understanding and as we accumulate more data. We might imagine a scenario in which we start with few data and little understanding. We first integrate existing knowledge via the systems modeling approach (systems analysis and the use of simulation models) to generate hypotheses about how the system works. This model-building process increases our understanding of the system and identifies specific areas in which important data are lacking. As we accumulate more relevant data, we use statistical analyses to interpret these data, thus generating further hypotheses concerning system structure and function that can be incorporated into our simulation models. Further simulation increases our understanding and identifies more data needs, and so on. Theoretically, as we continue to accumulate relevant data and increase our knowledge of the system, we eventually will achieve a complete understanding of the system and the definitive solution of our problem.

As ecological modelers, or as ecologists and natural resource managers who develop and use ecological models, we deal primarily with systems that are characterized by "organized complexity" for which we have relatively few data and for which we have little hope of ever accumulating a "complete" data set. That is, we find ourselves dealing with exactly those sorts of systems for which the systems modeling approach was developed. The systems modeling approach does not replace the other methods of problem solving but, rather, provides a framework that allows effective integration of the knowledge gained from the description, classification, and mathematical and statistical analysis of our observations of the world.

11.3 **Ecological modeling as a problem-solving process**

One of the greatest benefits of the modeling process as a problem-solving process within the contexts of ecological research, environmental decision-making, and environmental education is that it facilitates unambiguous communication among multidisciplinary, multivocational groups of people interested in a common problem. This benefit is well known (e.g., see Holling, 1978; Grant, 1998; Van den Belt, 2004), but arguably we do not take advantage of this benefit as often as we should. Within the context of ecological research, the modeling process facilitates integration of multidisciplinary teams of researchers and enhances research continuity by recording, in a dynamic way, via the series of intermediate developmental models, the history of the program. As scientists and graduate students join and leave the program, and as perceptions of individual investigators change, the evolving model provides a common point of reference for incorporation of new team members, and also provides the context within which past, present, and proposed projects can be integrated and evaluated.

Within the context of environmental decision-making, the modeling process facilitates multidisciplinary planning and creation of an effective communication interface between scientists, stakeholders, and policy makers. Here, we define "stakeholders" literally as anyone who has a vested interest in the system. Such a multivocational, multidisciplinary approach is a prerequisite for developing policies promoting sustainable use of our natural resources. In support of these claims, we offer two simple statements that were controversial 50 years ago, when ecological modeling was in its infancy, but now are accepted almost universally. (1) Environmental decision-making requires quantitative models. (2) Modeling is much too important to be left exclusively to modelers. We now recognize that virtually all ecological systems are affected either directly or indirectly by human activities. These impacts are being regulated increasingly by public authorities and policies. However, complexity of ecological systems and the extent of temporal and spatial scales of concern make it virtually impossible for field and laboratory studies to inform policy decisions adequately. Nonetheless, we still must make policy choices even when needed experimentation is impossible, and choices always are based on inferences (predictions) about possible outcomes of alternative policies. Since these inferences are unavoidable, we should make assumptions underlying our inferences as clear as possible. Precision (not to be confused with accuracy) inherent in quantitative representation of simulation models makes our hypothesized cause-effect relationships explicit, helps to identify our faulty logic and hidden assumptions, and helps to clarify differences of opinion that may exist among members of the modeling team with regard to appropriate representation of system structure and function.

To the extent that "nonmodelers," (stakeholders, policy makers, and other end-users of the model) become involved actively in the modeling process, they effectively become modelers. Thus, they reap the heuristic benefits associated with the modeling process that we described in the previous chapter. The model evolves into the communication interface among scientists, stakeholders, and decision-makers as team members develop a sense of joint ownership of the model. They also begin to speak and think in terms of "our" model rather than "their" model. The policy makers no longer view the model and model output as an attempt by modelers to preempt their decision-making prerogative, but rather as an additional source of information which they can factor into their decision-making process with full awareness of the inherent uncertainties. Stakeholders are more likely to use their votes wisely when electing public officials and choosing among propositions affecting the environment. Arguably, joint participation of scientists, stakeholders, and policy makers in the entire modeling process often does more to foster truly adaptive management of our natural resources than does detailed analysis of particular simulation results produced by the model.

Within the context of environmental education, the modeling process facilitates integrative thinking, active learning versus passive learning, and effective communication. The systems perspective provides the common conceptual framework and vocabulary needed to develop integrated educational programs that promote interdisciplinary communication and active intellectual teamwork within a problem-solving environment which facilitates effective learning. The process of developing a quantitative systems model requires precise representation of ideas concerning relationships that control dynamics of the system-of-interest. Whereas both ordinary spoken and written language allow ambiguous, incomplete, and even illogical statements to go unnoticed, simulation of dynamics of the system-of-interest demonstrates the numerical consequences of our assumptions and our logic, and identifies flaws in our understanding of causal relationships within the system. The opportunity for modelers to investigate rigorously and rapidly effects on system dynamics of altering assumptions and logic fosters active participation in the learning process.

11.4 Expectations of ecological models: Some thoughts for modelers

So if the usefulness of ecological modeling results as much from the process as from the final product, what expectations should we have of the final model? That is, what should we, as modelers, expect to learn from the simulation results we generate using our final model?

One expectation that we definitely should not have is that we will be able to predict (with absolute certainty) the future. Perhaps this seems a silly thing to mention. But the tendency to implicitly append "with absolute certainty" to "predict" is perhaps the most widespread and dangerous misconception about models, ecological or otherwise. Although we commonly phrase our model objectives in terms of "predicting" future behavior of our system-of-interest under different scenarios, what we really mean is that we will project, always with some degree of uncertainty, possible alternative futures of our system-of-interest under certain restrictive assumptions. Semantics aside, the point is that predicting/projecting/forecasting/ prophesying the future with complete certainty is not a legitimate expectation of our final model.

However, we should expect that the simulation results we generate with our final model will provide new knowledge that will help us to reduce, in a useful way, the uncertainty with which we view the future of our system-of-interest. That is, we should expect to gain knowledge from our simulations that will allow us to discard as highly unlikely some system futures that previously seemed quite plausible. The relative amount of knowledge that we gain depends in large part on the current level of our knowledge about the system-of-interest. Logically, the less we currently understand about the system, the more we are likely to learn. The fact that we are using the systems approach and simulation to solve our problem often implies that we are dealing with a system for which we have relatively few data and, quite likely, a relatively low level of understanding (Fig. 11.2), or, at least there are large holes in the available data base and large gaps in our understanding of system processes.

Related to the legitimate expectation that we will gain useful knowledge about our system-of-interest from our final model is the erroneous expectation that the amount of knowledge we gain will be directly related to model complexity. The amount of knowledge we gain is, indeed, related to model complexity, but the form of the relationship depends on our current level of knowledge about the system (Fig. 11.3). As we increase complexity in models of systems for which we have little knowledge, the relative amount of knowledge gained will increase rapidly at first, and relative gains will be great. Even the very general patterns of behavior generated by our simple models will add substantively to our knowledge. But soon, further increases in complexity of our models will yield little, and finally no, new knowledge. Benefits to performance of our models that are associated with explicitly representing more system components will be outweighed by detriments associated with the cumulative effect of errors inherent in estimation of an increasing number of model parameters. As we increase complexity in models of systems for which we have much knowledge, the relative amount of knowledge gained will increase slowly at first, and relative gains will be small. Most general patterns of system behavior generated by our simple models already will be a part of our knowledge base. However, as complexity continues to increase, the relative amount of knowledge gained will begin to increase more rapidly. Eventually, benefits to model performance will be outweighed by detriments, just as in systems for which we have little knowledge, but at a higher level of model complexity. By definition, the relative amount of knowledge gained will be less in the systems for which we already have much knowledge.

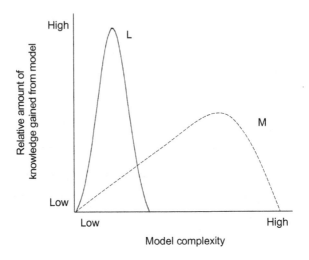

FIG. 11.3

General relationships between model complexity and relative amount of knowledge gained from the model in systems for which we currently have much (M) and little (L) knowledge.

Modified from Jørgensen, S.E., Bendoricchio, G., 2001. Fundamentals of Ecological Modelling.
Elsevier, Amsterdam, The Netherlands.

11.5 Expectations of ecological modeling: Some thoughts for environmental decision-makers

In the previous sections, we offered some reflections on usefulness of the modeling process itself. We considered how ecological modeling fits into the broader context of problem solving, focusing on the heuristic value to modelers of the ecological modeling process. In this section, we reflect on the heuristic value of ecological modeling from the perspective of model users, especially from the perspective of those using models as an aid in environmental decision-making. At first glance, it may seem natural to assume that the heuristic value of modeling accrues primarily to model developers. This may be true if decision-makers use the final model in a "black-box" mode. That is, if simulation results are interpreted at face value, without a basic understanding of the processes within the model that generated those results, and without an appreciation of the uncertainty associated with model predictions. However, ideally, and ever more frequently in reality, end-users of a model are involved actively in the modeling process and thus share in the heuristic value of modeling. Through a dialog with model developers, end-users gain a basic understanding of the processes within the model that are generating the forecasts in which they are interested. Even more importantly, they gain an appreciation of the uncertainty associated with model predictions. Below, we suggest some expectations of ecological modeling which, ideally, are held by "enlightened" environmental decision-makers.

11.5.1 **Ecological modeling is a scientific process**

Ecological modeling should be viewed as a scientific process that seeks to simplify reality to enhance our understanding through structuring our data, knowledge, and assumptions in a disciplined way for a specific purpose. Our observational data and existing knowledge rarely are sufficient for establishing causal relationships in complex natural systems. At best, the data tend to be too limited in scope and our knowledge tends to be too compartmentalized for this purpose. Models can integrate and capture systematically our understanding of how changes in climate, demographics, management, and other factors affect selected indicators of system dynamics so that consequences of management options can be clarified. The more complex the system being considered, the greater the role for models to account for interactions among system drivers, processes, and outcomes. On the other hand, models are imperfect representations of reality, and the nature of the systems they attempt to describe almost always changes over time. But new knowledge and data acquired about the system, appropriately targeted, almost always facilitate modifications to the model that benefit its usefulness.

11.5.2 **Ecological modeling is an art form**

Ecological modeling should be viewed as an art form in which two fundamental phases of thinking alternate with one another in a sort of adaptive dance between induction and deduction (Walters, 1986). There is an inductive, creative, synthetic, artistic, constructive phase when we try to decide what to include in this "caricature" of reality which is the model, and how to include it. Then comes a deductive, more mechanical phase when we use mathematical analysis and simulation, employing the caricature as a "deductive engine." We then compare the deductions with our expectations and, if inconsistencies are revealed, the dance continues with another round of induction. Learning is involved in both phases. We may be surprised equally by the gaps in our understanding that attempts at synthesis usually reveal, and by the predictions themselves. Thus, ecological modeling, and modeling in general, should be viewed as a very human way of searching for understanding.

11.5.3 **Ecological modeling is a compliment to, but not a replacement of, field experience**

Ecological modeling should be viewed as is a compliment to field experience. That modeling, ecological or otherwise, is a substitute for field experience is a myth. The often-heard claim that models can be used as "laboratory worlds" to "test" possible effects of management options that are too big or too expensive or too dangerous to study experimentally has a useful connotation, but should not be taken literally. Of course, there can be no test, in the scientific sense, when experimentation or observation is impossible. A more precise statement of the "laboratory world" concept requires some cumbersome verbiage, but might be stated, paraphrasing

Walters (1986), as follows. We still must make management decisions even when experimentation is impossible. The decisions always are based on some sort of inference about alternative outcomes. Since inference (i.e., prediction) is unavoidable, we should make assumptions underlying the inference (i.e., the laboratory world) as clear as possible, if for no other reason than to avoid mistakes in our reasoning (our hidden assumptions, our incorrect deductions). In other words, since modeling in some general sense is unavoidable, we should make the modeling process as transparent as possible.

11.5.4 Ecological modeling is an integral part of adaptive environmental management

Ecological modeling should be viewed as an integral part of adaptive environmental management. The challenge is to view and implement modeling as an ongoing scientific and participatory process that serves adaptive management. In this connection, the first aim of modeling should be to identify new knowledge and data needs that will lead to further understanding, if not to direct clarification, of the impacts of various courses of action on system dynamics. Another aim should be to assist in obtaining consensus on, or defensibility of, management decisions to be taken. Such aims demand transparency in the modeling practice, especially in the selection of a model type and approach, which recognizes the context in which the model will be used. Such aims also call for use of the participatory processes and analytic tools that will help identify the types of data, knowledge, and associated experiments that will facilitate integration of the modeling process within the particular environmental management context at hand.

11.5.5 Ecological modeling "embraces" uncertainty

Ecological modeling should be viewed as a way of "embracing" environmental uncertainty. As we mentioned in the opening section of this chapter, we must form our expectations of ecological models in view of the uncertainties associated with their development. Modeling helps to clarify and to highlight uncertainties about structures and functional relationships within the system under study, but usually we cannot resolve all these uncertainties via modeling, or, for that matter, via any other problem-solving methodology. Thus, most environmental decisions, whether or not they are based on models, speaking quite frankly, are essentially gambles. Although it may be uncomfortable to think about environmental decision-making as gambling, the analogy is useful. Statistical decision theorists suggest that the way to embrace uncertainty is to assign odds or probabilities to alternative models, and use these odds in decision-making to weight the relative importance of information gained from alternative models. In the next chapter, we will reflect in a more general way on how we might "embrace" uncertainty.

Embracing uncertainty

12

Chapter outline

12.1 Overview

Uncertainty is a pervasive problem in the development of ecological models precisely because uncertainty is a pervasive problem in ecology. As we have seen, uncertainty pervades the process of model evaluation. The broad array of qualitative and quantitative aspects of model structure and behavior that we examined in Chapter 6 as part of model evaluation all are focused implicitly on identifying and confronting our uncertainties regarding usefulness of our model. Sensitivity analysis, in particular, helps us determine the level of confidence that we should have in model projections in view of the uncertainty with which we have estimated those parameters that most affect model behavior. This parametric uncertainty often results from a lack of data with which to estimate model parameters. However, as we have seen, even if data from the real system are available, they are not an infallible standard for judging performance of the model. Thus, evaluation of uncertainty associated with field or laboratory data played a prominent role in both of the methodological approaches to model evaluation that we discussed in the previous chapter (i.e., Rykiel, 1996; Augusiak et al., 2014). Also from our discussion in Chapter 10, it is clear that another source of uncertainty associated with model evaluation arises from the semantics associated with discussions of the topic. This linguistic uncertainty arguably plays an under-appreciated role in the model development process.

In this chapter, we reflect upon various aspects of uncertainty associated with ecological model development. Since model evaluation is concurrent with the entire modeling process, uncertainty confronts us repeatedly at every step of way. That is, as we pointed out in our candid view of the modeling process in Chapter 9, we usually cycle through the four phases of the modeling process several times. We seldom quantify the entire conceptual model before running simulations and evaluating model behavior. Rather, we usually construct a simple "running" model as quickly as possible and then expand it gradually through a series of small additions until we have quantified the entire model (see Fig. 9.1). We evaluate each addition to the model carefully with regard to its usefulness in view of uncertainties associated with its inclusion.

Aspects of uncertainty relevant to ecological model development have been classified in a variety of ways within different ecological and environmental contexts. Of course, uncertainty and associated terms such as error, risk, and ignorance have been defined and interpreted by a wide variety of scientific authors outside the context of ecological modeling. Different terms and definitions associated with the concept of uncertainty that can be found in the literature reflect underlying philosophical differences among scientific disciplines, as well as differences among specific topics under consideration within the difference disciplines (e.g., see Ascough et al. (2008) for a review of uncertainty typologies presented in the literature from 1990 to 2008). The reflections that we offer in this chapter have been influenced heavily by the works of Regan et al. (2002), Walker et al. (2003), and Ascough et al. (2008), which we believe touch upon the primary aspects of uncertainty that merit our attention as we develop ecological models. These authors presented classification schemes ("taxonomies," "typologies") for various sources and types of uncertainty, which differ in detail but, we believe, overlap broadly in concept. Below, we first summarize briefly the classification schemes of these authors. We then offer a synthesis/reorganization of their ideas that we have found useful in thinking about ways to "embrace" uncertainty as we develop our ecological models.

12.2 **A brief summary of three uncertainty typologies**

Regan et al. (2002) classified sources of uncertainty within the context of ecology and conservation biology into two main categories: (1) epistemic uncertainty and (2) linguistic uncertainty. Epistemic uncertainty is uncertainty associated with knowledge of the state of a system. This includes uncertainty due to limitation of measurement devices, insufficiency of data, extrapolations and interpolations of data, and variability over time or space inherent in the system. Linguistic uncertainty is the uncertainty associated with our natural language, including our scientific vocabulary, much of which is ambiguous, vague, context dependent, underspecified, or exhibits theoretical indeterminacies.

Walker et al. (2003) classified sources of uncertainty within the context of model-based decision support, focusing on uncertainty perceived from the point of view of modelers. These authors discriminated among three dimensions of uncertainty: location of the uncertainty, level of the uncertainty, and nature of the uncertainty. Within location of uncertainty, they identified context uncertainty, model uncertainty, uncertainty related to model inputs, parameter uncertainty, and model outcome uncertainty. Within level of uncertainty, they identified statistical uncertainty, scenario uncertainty, uncertainty due to recognized ignorance, and uncertainty due to total ignorance. Within nature of uncertainty, they identified epistemic uncertainty and variability uncertainty. Epistemic uncertainty is due to our imperfect knowledge. Variability uncertainty is due to inherent system uncertainty.

Ascough et al. (2008) classified sources of uncertainty in (model-aided) environmental and ecological decision-making into three main categories: (1) knowledge uncertainty, (2) variability uncertainty, and (3) linguistic uncertainty, all of which contribute to decision uncertainty. Within knowledge uncertainty, they identified uncertainty associated with process understanding and three sources of model uncertainty: (1) parametric/data uncertainty, (2) structure uncertainty, and (3) technical uncertainty. Within variability uncertainty, they identified (1) natural sources, (2) human sources, (3) institutional sources, and (4) technological sources. Within linguistic uncertainty, they identified (1) vagueness, (2) ambiguity, and (3) under-specificity.

12.3 **A synthesis of uncertainty concepts for ecological modeling**

We have found it useful to think about uncertainty within the context of ecological modeling in terms of (1) level of uncertainty, (2) type of uncertainty, and (3) source of uncertainty. A simple dictionary definition of uncertainty is "a lack of sureness about something." Synonyms include doubt and skepticism. Note that the level of uncertainty may range from falling just short of certainty to an almost complete lack of knowledge. Also note that some types of uncertainty are reducible by gaining more knowledge about the system-of-interest, but some types of uncertainty are irreducible. The former has been referred to as "knowledge uncertainty" and the latter as "variability uncertainty." The ability to gain more relevant knowledge, of course, depends on the source of the uncertainty.

12.4 Levels of uncertainty

As we alluded to above, "uncertainty" encompasses a continuum of different levels of knowledge, or more precisely the lack thereof, ranging from the theoretical ideal of complete deterministic understanding at one extreme to total ignorance at the other extreme. As we move from the unachievable ideal of complete deterministic understanding toward total ignorance, we pass through three zones that Walker et al. (2003) labeled "statistical uncertainty," "scenario uncertainty," and "recognized ignorance." Lines separating these zones are fuzzy and the scale is arbitrary, not unlike the distinction made in basic ecology textbooks between "r-selected" species and "K-selected" species. But, like r- and K-selected species, distinguishing characteristics are clear, and relative position along the continuum is of particular importance when considering model projections of potential future states of the system-of-interest.

12.4.1 Statistical uncertainty

Statistical uncertainty refers to any uncertainty that can be described adequately in statistical terms. Statistical uncertainty can apply to any aspect of a model, even to aspects of model structure, as long as deviation from the true value can be characterized statistically. Statistical uncertainty is what usually is referred to as "uncertainty" in the natural sciences. However, an exclusive focus on statistical uncertainty implicitly assumes that functional relationships in the given model are reasonably good descriptions of phenomena being simulated, and that data used to calibrate the model are representative of circumstances to which the model will be applied. If this is not the case, "deeper" forms of uncertainty supersede the importance of statistical uncertainty in the uncertainty analysis.

Perhaps the most obvious example of statistical uncertainty is the measurement uncertainty associated with all data. Measurement uncertainty results from the fact that measurements never can represent precisely the "true" value of what is being measured. Measurement uncertainty in data can be due to sampling error or inaccuracy or imprecision in the measurements themselves.

12.4.2 Scenario uncertainty

Scenario uncertainty refers to any uncertainty associated with the choice of scenarios simulated by the model. A scenario is a plausible description of how the system and/or the forces that drive the system may develop in the future. Thus, scenario uncertainty results from uncertainty related to the future external environment of the system-of-interest, often its future physical environment, and effects of that environment on behavior of the system-of-interest. To be plausible, a scenario should be based on a coherent and an internally consistent set of assumptions about key functional relationships and driving variables within the model. However, since scenario development implies making of assumptions that in most cases are not verifiable, scenario uncertainty is associated with uncertainty at a level beyond statistical uncertainty.

The demarcation in the transition from statistical uncertainty to scenario uncertainty is at the point where a change occurs from a consistent continuum of outcomes expressed stochastically to a range of discrete possibilities where choices must be made with respect to options to be analyzed without reference to their statistical likelihoods. With statistical uncertainty, functional relationships are described well and a statistical expression of the uncertainty present can be formulated. With scenario uncertainty, there is a range of possible outcomes, but mechanisms leading to those outcomes are not well understood, and it is not possible to formulate the probability of any one particular outcome occurring.

12.4.3 Recognized ignorance

Recognized ignorance refers to fundamental uncertainty about processes operating within the system-of-interest. We know neither the functional relationships that generate patterns of system behavior nor the statistical properties of those resulting patterns. The uncertainty due to recognized ignorance can be divided further into reducible ignorance and irreducible ignorance. Reducible ignorance can be resolved by conducting further research. In this regard, reducible ignorance is analogous to knowledge uncertainty. Irreducible ignorance cannot be resolved by any means. Irreducible ignorance is referred to as indeterminacy, which is distinct from variability uncertainty.

12.4.4 Total ignorance

Total ignorance refers to the other extreme from determinism on the scale of uncertainty, which implies a deep level of uncertainty. In fact, we do not even know that we do not know and, thus, we have no way of knowing the full extent of our ignorance.

12.5 Types of uncertainty and their sources

We find it useful to distinguish between the two types of uncertainty, knowledge uncertainty and variability uncertainty, in terms of their potential reducibility. Both types of uncertainty can exist at any point along the uncertainty level continuum. Knowledge uncertainty, also known as epistemic uncertainty or epistemological uncertainty, refers to uncertainty due to the limitation of our knowledge. Variability uncertainty, also known as external uncertainty, random uncertainty, or stochastic uncertainty, refers to uncertainty due to the inherent variability manifested in natural and human systems. Theoretically, knowledge uncertainty can be reduced by additional research and empirical observations, but variability uncertainty cannot be reduced. In practice, it is difficult to identify precisely those sources of uncertainty that are reducible through investigations and research versus those sources of uncertainty that are irreducible because of their inherent variability.

A particular source of a given uncertainty might be categorized appropriately as knowledge uncertainty within one problem context at a particular point in time and as variability uncertainty within a different problem context, or within the same problem context at a different point in time. Ideally, sources of uncertainty should be characterized in a way that is operationally beneficial to understanding where in the entire modeling process uncertainty associated with the model outputs is generated.

Also, uncertainty associated with a certain process or a certain event might include both knowledge uncertainty and variability uncertainty. An example is the uncertainty associated with a 100-year flood event at a particular site. This flood event can be estimated, for example, by use of a standard flood frequency analysis based on existing flow data. Knowledge uncertainty may be reduced by improving the data analysis, by making additional monitoring (a longer time series), or by deepening our understanding of how the system-of-interest works. However, no matter how perfect both the data collection and the mechanistic understanding of the system are, and no matter how long the historical data time series is, there always will be some variability uncertainty inherent in the natural system related to the stochastic and chaotic nature of natural phenomena such as weather. Even perfect knowledge of these phenomena could not give us a deterministic prediction of future floods, but, rather, would give us the form of a perfect characterization of the natural variability.

Below, for expository purposes, we have associated various general sources of uncertainty encountered during the modeling process with either knowledge uncertainty or variability uncertainty. However, our intent is not to propose a specific categorization scheme, but, rather, to emphasize the importance of making an explicit effort to identify, estimate if possible, and prioritize in a systematic manner all of the aspects of a modeling project that contribute to the uncertainty associated with model outcomes.

12.5.1 Sources of knowledge uncertainty

Model context uncertainty

Model context uncertainty refers to uncertainty regarding identification of boundaries of the system-of-interest, that is, identification of components of the real system that should be included within the system-of-interest. This source of uncertainty also refers to uncertainty regarding issues to be addressed by the model and selection of outcomes of interest to be estimated by the model. It is important to note that the most useful model may not be the most complex or "complete" in the sense of explicitly representing every aspect of the system under study. Representation in the model of many details actually may increase uncertainty regarding the most important cause-effect relationships that generate the most relevant model outcomes.

System process uncertainty

Process uncertainty refers to uncertainty associated with lack of scientific understanding of processes that generate behavior of the system-of-interest. This may take the form of an inability to identify the processes involved, an inability to understand even qualitatively how the processes interact, or an inability to quantify the processes appropriately. Process uncertainty also may come in the form of inability to identify appropriate temporal or spatial scales at which to represent processes. It is important to note that new knowledge on complex processes may reveal the presence of uncertainties that were previously unknown or understated (Walker et al., 2003). In this way, somewhat ironically, additional knowledge may reveal that our understanding is more limited or that processes are more complex than we previously thought (van der Sluijs, 1997).

Model structure uncertainty

Model structure uncertainty refers to uncertainty regarding the amount of detail with which model components should be represented and the form of relationships that should be used to describe connections among model components. This source of uncertainty also refers to uncertainty regarding choices of surrogate variables and decisions to exclude variables.

Data uncertainty

Data uncertainty refers to uncertainty resulting from data measurement errors, systematic errors in data collection, and subjective interpretations of data. Data measurement errors may arise due to types of instruments used, quality and frequency of instrument calibration, data reading/logging errors, biases in data sampling or data retrieval, and errors in data transmission/storage. This source of uncertainty also may arise from types of data recorded, types of data analysis/processing, and methods of data presentation (Maier et al., 2008). Systematic errors in data collection may arise due to biases in measuring equipment or sampling procedure, from consistently incorrect recording of measurements, and from deliberate judgment of scientists to exclude (or include) data that ought not be excluded (or included). Uncertainty due to subjective interpretations of data arises most often when data are scarce and error prone.

Model parametric uncertainty

Model parametric uncertainty refers to uncertainty associated with estimates of model parameters. Since model parameter estimates generally have to be obtained directly from measured data or indirectly from measured input-output data by calibration, there is a strong relationship between parametric uncertainty and data uncertainty. This source of uncertainty may result from estimation of parameters through nonrepresentative samples caused by temporal, spatial, or financial limitations associated with data collection. It also may result from calibration of too many model parameters. In fact, as a general rule, parametric uncertainty increases with increases in number of parameters calibrated.

Model technical uncertainty

Model technical uncertainty refers to uncertainty associated with computer implementation of the model. This includes uncertainty generated by hardware or software errors, that is, hidden flaws in technical equipment. Hardware errors arise from flaws in hardware components such as processors. For example, it is possible for the number of significant figures represented in floating-point calculations to affect model outcomes when sensitivity to initial conditions is an important factor, such as in chaotic systems. Software errors arise from flaws in software, design errors in algorithms, and typing errors in model source code. For example, regarding design errors in algorithms, uncertainty may arise when continuous equations are used to describe discrete processes or vice versa. Such uncertainty is inherent in mathematical expressions used to represent the physical and biological world.

Model output uncertainty

Model output uncertainty refers to accumulated uncertainty associated with model outputs, which results from propagation through the model of all sources of uncertainty. This sometimes is called prediction error, since it represents uncertainty regarding discrepancy between the true values of outcomes in the real world and the values predicted by the model. If true values are known, then a formal validation exercise can establish the prediction error. However, practically all simulation models are used to extrapolate beyond known situations to estimate outcomes for situations that do not yet exist. Thus, the discrepancy remains uncertain.

12.5.2 **Sources of variability uncertainty**

Natural variability uncertainty

Natural variability uncertainty refers to the inherent randomness of nature, that is, the unpredictable quality of natural processes. This uncertainty is reflected in uncertainty associated with external input data, input functions, and empirically based estimates of model parameters. Many measurable properties of real-world systems vary over space and/or time in a manner that is beyond control, simply due to the nature of the phenomena involved. Some taxonomies of uncertainty distinguish between temporal and spatial components of natural variability uncertainty (e.g., Chesson, 1978). A point worth noting here is that, although many real-world processes are said to be inherently random, genuine examples of this kind of uncertainty are rare. We perceive some processes as being random because we do not have enough information about the dynamic systems in which they occur to make any sensible estimates about their outcomes. Such processes are for all practical purposes inherently random, but they are not genuinely inherently random.

Human behavioral uncertainty

Human behavioral uncertainty refers to uncertainty resulting from discrepancies between what people say they will do and what they actually do. This includes uncertainty associated with nonrational behavior and, more generally, with any deviation from "standard" behavioral patterns of individuals. Human behavioral uncertainty has received relatively little attention in the ecological modeling literature. However, this type of uncertainty can have a significant impact on the modeling process. Not surprisingly, results of a modeling exercise almost always vary significantly when different modelers develop separate models based on the same problem and the same data. That is, knowledge, experience, and preferences of modelers significantly impact modeling outcomes (Refsgaard et al., 2005). The knowledge base, education, attitudes, and political "clout" of stakeholder groups also can have a major influence on the modeling process, particularly with regard to choice of scenarios to be simulated and choice and weighting of scenario assessment criteria to be employed. But even more technical aspects of the modeling process are not immune from uncertainty resulting from human behavior.

Societal uncertainty

Societal uncertainty refers to uncertainty resulting from the unpredictable nature of societal processes. This includes uncertainty associated with social, economic, institutional, and cultural dynamics. Although bioeconomic modeling has a long history, explicit representation of societal, institutional, and cultural processes in ecological models, arguably, still is in its infancy.

Technological surprise uncertainty

Technological surprise uncertainty refers to uncertainty resulting from new breakthroughs in technology or unexpected consequences ("side-effects") of existing technologies. Initially, these sources usually are categorized as variability uncertainty, but it may be difficult to identify precisely what is reducible through investigations and research, and what is irreducible because it is an inherent property of the technology.

12.5.3 What about linguistic uncertainty?

Both Regan et al. (2002) and Ascough et al. (2008) also mentioned linguistic uncertainty as a major type of uncertainty. Both groups of authors identified vagueness, ambiguity, and under-specificity as sources of linguistic uncertainty, with Regan et al. (2002) adding context dependency and indeterminacy of theoretical terms as an additional two sources. All of these uncertainties arise in natural and scientific language, and all can impact the modeling process. Of these, vagueness arguably is the most important for practical purposes. We recall our discussion of semantics related to model evaluation in Chapter 10. Both Regan et al. (2002) and Ascough et al. (2008) classified linguistic uncertainty separately from knowledge (epistemic) uncertainty and emphasized its importance. We agree that communication skills unquestionably are critical to success of a modeling project, but would refer readers to these two articles and the literature cited therein for further discussion of this topic.

Integration of existing models

13

Chapter outline

13.1 Overview

In this chapter, we reflect upon the practice of integrating existing models, rather than developing our own model, to address our objective. This is a practice that has existed since the earliest days of ecological modeling, but is a practice that is expanding rapidly with the aid of ever-accelerating advances in computer technology. We hasten to note the distinction between what we will refer to as "integrated" models and what we will refer to as "integral" models. Integrated models, which are the focus of this chapter, result from integration (quantitative coupling) of models that originally were developed independently of one another. Integral models, for the purposes of avoiding confusion in this chapter, we define as models that are custom-built from "scratch," which may or may not consist of integrated, multidisciplinary components. That is, if they do consist of multidisciplinary components, integration begins during formulation of the conceptual model.

We have focused this book on development, evaluation, and application of ecological models that are custom-built from "scratch," as opposed to the integration of existing models. That is, we begin with a "blank slate" and custom-build our model to address a specific objective. In particular, our emphasis has been on development of explanatory ("mechanistic") simulation models (see Chapter 1, Section 1.2.4 for definitions of terms). The model objective guides every phase of model development. We begin with conceptual model formulation, and then proceed to quantitative model specification, model evaluation, and model application (Fig. 3.1). We begin conceptual model formulation with a clear statement of model objectives, and then proceed to bound the system-of-interest, categorize components within the system-of-interest, identify relationships among the components of interest, represent the conceptual model diagrammatically, and describe expected patterns of model behavior (Fig. 4.1). In the very first section of this book, we stated that the usefulness of ecological simulation modeling often results as much from this process of model development as from the final model and model application simulations (Chapter 1, Section 1.1). We noted that an appreciation of the heuristic aspect of the modeling process, as well as development of modeling skills, is gained primarily through practicing this process. We also cautioned that we must maintain a keen awareness of exactly what we are doing (in practice), why we are doing it (in theory), and why it makes (common) sense. Without this awareness, we risk making silly mistakes, which often come hidden under a thick covering of sophisticated quantitative techniques and the associated jargon, to which not even the most experienced modelers are immune. Given our primary focus and these precautionary statements, one might infer that an implicit argument against the growing practice of integrating existing models permeates this book. This is not our intention. Rather, as we hope this chapter will make clear, our intent is to call attention to the increased risk of underestimating the importance of the first phase of the modeling process, conceptual model formulation, that lurks within the process of the integrating existing models.

In the following sections of this chapter, we first provide the rationale for integrating existing models. We next describe the general process involved in coupling already built models within a single computing framework. We then suggest some potential dangers of linking together complicated extant models which lurk behind sophisticated technological advances in computing hardware and software, and which may be more easily overlooked than is the case when dealing with simpler stand-alone models. Finally, we offer some thoughts concerning the future of integrated models.

13.2 Rationale for integrating existing models

The rationale for integrating existing models might be described most simply by the phrase "there is no need to recreate the wheel." Extending this analogy just a bit further, the number and variety of wheels available is huge and is expanding exponentially. Within the context of ecological and environmental modeling, a more sophisticated expression of the rationale is that we need to integrate information from a variety of scientific fields, including diverse types of data, theories, and models, to address the complex, multidisciplinary problems that confront us in today's world.

Our systems-of-interest often consist of complex physical environmental, ecological, economic, and social components, which are interrelated with one another. The use of integrated approaches has become a common theme in environmental management, and use of quantitative models to support the environmental decision-making process is widespread. Increasingly, environmental management requires multiscale, multiobjective assessment and decision making which considers economic and social systems, as well as the ecosystem. There are several widely accepted and readily available atmospheric (e.g., Hybrid Single-Particle Lagrangian Integrated Trajectory Model—HYSPLIT) (Stein et al., 2015), hydrologic (e.g., MODFLOW) (Hughes et al., 2017; Langevin et al., 2017), and oceanographic models (e.g., Modular Ocean Model—MOM) (Griffies, 2012) that provide sophisticated representations of general physical environmental processes. There also are numerous well-known "legacy" models that contain ecological components, which have been developed over decades (e.g., IBP grassland model which dates back to PAWNEE, GAP models, others). Thus, it is logical to explore possibilities of integrating these established models to address our multidisciplinary ecological and environmental questions, which are embedded in complex systems. Such integration efforts constitute an active and growing area of research whose primary goal is to facilitate increased reuse and interoperability of science-related software components contained in different models.

13.3 General process of model integration

The general process of model integration might be summarized as consisting of five phases: (1) preintegration assessment, (2) preparation of models for integration, (3) orchestration of models during simulation, (4) data interoperability, and (5) testing integration (Belete et al., 2017) (Fig. 13.1). The basic challenge is to facilitate linkage of models that were built originally to stand alone using different tools, methodologies, and assumptions within a single computing framework. That is, to make stand-alone models discoverable, reusable, and interoperable (Goodall et al., 2011). Discoverability refers to availability of meta information of a resource such that it can be searched for and located efficiently, and understood to ensure proper use (Erl, 2008). Reusability refers to how existing software can, in whole or in part, be used to develop new software (Frakes and Kang, 2005), as well as the degree to which a module of code can be used in more than one system or in building other modules of code (ISO/IEC, 2011). Reuse also implies a decision process that includes evaluating software for its ability to serve a new purpose. Interoperability refers to ability of a system or a product to work with other systems or products without special effort by the user (Chang and Lee, 2004). In the following sections, we describe each of the five phases in the general process of model integration. We have drawn heavily on the excellent recent review of the model integration process by Belete et al. (2017) for much of this material.

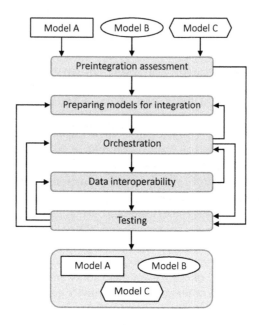

FIG. 13.1

The general process of model integration consisting of five phases: (1) preintegration assessment, (2) preparation of models for integration, (3) orchestration of models during simulation, (4) data interoperability, and (5) testing.

Modified from Belete, G.F., Voinov, A., et al., 2017. An overview of the model integration process: from pre-integration assessment to testing. Environ. Model. Softw. 87, 49–63.

13.3.1 Preintegration assessment

Preintegration assessment combines science-based requirements related to the objective of the modeling project with computer technologies needed to create an initial integrated software system workflow design. Preintegration assessment represents a key aspect of model integration since it establishes roles and relationships among scientists who are addressing the project objective and software engineers who are writing computer code to automate the integration process. The connection between software integration frameworks and science domains cannot and should not be ignored.

On the science side, preintegration assessment includes a problem statement, articulation of purpose and goals of the science-based modeling effort, conceptualization of the relevant system (including major system components and interactions), characterization of scenarios to be simulated, analysis of alternative scenarios, design of a science-based approach, and a statement of any project resource constraints and solution criteria. Collectively, this information represents requirements that must be satisfied by either an existing or a new software system.

The main cost of model integration is the effort required to detect and resolve inconsistencies in the coupled system (Ramamoorthy et al., 1992). This cost can be reduced if inconsistencies are detected and resolved prior to model integration. That is, during the process of transforming the science design into a software design. Procedures that aid this transformation include development of context views that define relationships, dependencies, and interactions between the system and its environment; use cases that list interactions between the system and the user; activity diagrams that show the flow of activities while using the system; and diagrams that provide a high-level description of system behavior; all of which are recommended by the system-of-systems engineering approach (Butterfield et al., 2008).

Belete et al. (2017) identified 12 sets of questions, the answers to which facilitate pre-integration assessment. (1) Why are we integrating? What are the objectives? (2) What system is to be simulated? What scenarios are to be simulated using the integrated system? (3) What indicator variables are to be tested using the integrated system? What are the options for calculation of the indicator values? (4) Are required modeling components from the same science domain? Are the semantics compatible? Are there frameworks that cover the relevant science domains, or will you need to fit into a "foreign" framework or develop both the new framework and models within this framework? (5) How much data and information will the models exchange, for example, one variable versus hundreds of variables, how often, and over what spatial grid and temporal resolution will data by exchanged? What processes require feedback loops? (6) Is it important (or necessary) to keep participating components independent of one another, or will linking be established by creating a permanent dependency of one component upon another, effectively merging multiple components into a single component? (7) What are the implications of programming languages, operating systems, and computer platforms used? Will language interoperability tools be required? Will individual components remain on their respective platforms (e.g., development environment of operating system) or will they be migrated to a common platform in the integrated system? (8) What programming skills are required and are they available? How much new software will be required to link the models? (9) How will system input data (i.e., data about the physical/chemical/biological/social system being modeled) be organized and distributed among components? (10) Are data unit conversion functions needed? (11) Is dataset translation, aggregation/disaggregation required? If so, are the necessary tools available or must they be created? (12) Are there specific performance requirements to satisfy (e.g., parallel computation)? Consideration of these 12 conceptual and technical questions forms the basis for an initial design that will enable evaluation of feasibility of the integration and resource requirements. If model integration is deemed feasible, the process continues with preparation of the models for integration.

13.3.2 Preparation of models for integration

From a software engineering perspective, preparing a component for integration depends on its native software environment (language and operating system), the manner in which inputs and outputs are organized (e.g., using files, databases, and so on), and the integration approach used by the target system (framework). Belete et al. (2017) offer the following example to illustrate the situation. Suppose we have a model that produces climate change scenarios up to the year 2100. The simulation start year is constant, but the end year is a user input. The model's output, as originally designed, is accessible only after the final time step. If we want to link this model with another model so that data is exchanged at the end of each time step, we must make certain modifications. These modifications include preparing a function that returns values of required variables at the end of each time step, and a function that receives values of selected variables from the model to be linked. In achieving this linkage, we need to consider that the two models may be developed using different languages, may be deployed on different platforms, may treat space and time differently, and so on. Thus, the process of preparing models for integration can go through a variety of combinations of steps.

Model modification sometimes will be needed to enable automated input-output data exchange that conforms to a specific framework design. In such situations, small existing (legacy) models can be reprogrammed, or we may directly modify the model interface to conform to the framework data exchange protocol that is required. For larger and more complex models, however, rewriting is time-consuming and error-prone (Peckham et al., 2013). To avoid rewriting, it is common to develop a communication wrapper (Van Ittersum et al., 2008) which is a thin layer of code on top of the native component that manages data exchanges between the native component and the other components or the system. Wrappers could be developed to make a model language-interoperable, make a model accessible over a network, implement agreed-upon names for functions and variables, provide a custom interface with required input-output attributes, or address a combination of these objectives. Depending on the integration framework, a model wrapper can be developed following industry standards or local standards. Generally, the work required to develop wrappers depends on how much we want to adapt the existing model interface.

13.3.3 Orchestration of models during simulation

A complementary part of technical integration is orchestrating models during simulation, which requires enabling and managing interactions between components (Madni and Sievers, 2014). The orchestration task includes: identifying components to be included, establishing linages and overall workflow to be implemented, and managing execution of the workflow (initializing components, synchronizing execution, facilitating data exchange, error management, and system file management). Furthermore, the operational nature of the data exchange within an integrated system may be sequential (feed forward) or iterative (feedback). In feed forward systems, one component completes its execution and produces data for subsequent consumption. In feedback systems, components require the exchange of intermediate data values at certain time steps.

The actual methods employed for orchestrating the technical interoperability of models vary among development groups. Belete et al. (2017) have categorized these methods as component-based, web service-based, hybrids of component-based and web service-based, and tailor-made (custom-made) techniques, and cited examples of each.

13.3.4 **Data interoperability**

Data interoperability is the ability to correctly interpret data that crosses system or organizational boundaries (Renner, 2001). At the core of integrated modeling simulations is the exchange of data between workflow components. Implicit in discussions of component preparation and workflow orchestration aspects of model integration is the need to ensure that the data being transferred between components is both semantically interoperable and structurally interoperable. Semantically interoperable means that each dataset has unambiguous meaning and is correctly mapped and, if necessary, translated. Structurally interoperable means that the dataset is formatted in the required from. Semantic interoperability issues for data are the largest hurdles in model integration (Argent, 2004). Differences in the naming of entities, the scales used to represent space and time, and the ways we represent concepts in relationship to one another are the most common causes of semantic heterogeneity among models.

It often is necessary to perform additional computations on data as it is transferred from one model to another. Data mediation requires identification of the WHAT, WHERE, and WHEN of data within the context of each model (Moore and Tindall, 2005). WHAT is the name, definition, and quantitative description of the data represented by the variable, including factors such as units of measure, and allowable minimum values and maximum values. WHERE is location information and spatial resolution. WHEN describes temporal information and resolution. While space and time are obvious contexts, other "dimensions" must be considered as well. For example, differences between how components categorize chemicals, land use, or human/ecological species demographics must be resolved. Thus, unlike technical integration of modules, developing a "generic" module that will handle data mediation among different models is quite difficult.

A related issue involves determining which component will perform the data mediation. An independent module can be dedicated to this purpose, or the particular modules themselves can mediate that data. In most integration frameworks, a dedicated module for data interoperability is developed, but participating models also can be responsible.

Frameworks generally vary in the degree to which they codify the metadata and conversions related to data exchange. Data interoperability is achieved either by hard-coding, using an in-house developed standard and annotations, or via a controlled vocabulary and ontology. Belete et al. (2017) describe various frameworks for data interoperability in current use.

13.3.5 Testing integration

Testing evaluates whether a system satisfies certain specified requirements and improves the integrated system's quality. Integration of models is error-prone since it requires both scientific efforts and programming efforts (Bruggeman and Bolding, 2014). Even though individual components are tested independently and may perform correctly, integration output still can produce unexpected results due to different assumptions made during integration (Voinov and Cerco, 2010). Testing remains difficult regardless of the integration strategy used (Ramamurthy et al., 1992). There are many potential sources of error in designing and implementing software to facilitate data exchange within integrated modeling systems, including matching of variables, conversion of units of measure, use of metadata to perform semantic operations, and translation of space and time dimensions. Identifying and reporting specific sources of error (debugging) in integrated systems also is a challenge.

Integration of complex data and models is likely to propagate uncertainty throughout the model chain (Dubois et al., 2013) and uncertainty tends to increase as more models are chained together (Geller and Turner, 2007). Quantifying and communicating uncertainty in a model chain is challenging but crucial if we want to ensure that the integration output will be accepted by the larger user community. In addition, time-stepping and numeric integration methods selected for linked components have their own effect on error and uncertainty in integration output (Belete and Voinov, 2016).

From the perspective of software engineering, testing occurs at different stages of development including unit testing, integration testing, system testing, and acceptance testing (Luo, 2001). Unit testing is the lowest level and is performed to verify correct component function. Integration testing assesses whether interfaces between different components of a system are communicating correctly. System testing attempts to check end-to-end functioning of the system, and acceptance testing is done to deliver the system to end users. We also must consider emergent properties that appear when we link components into integrated systems (Ammann and Offutt, 2008). Best practices indicate software testing can be improved by characterizing the Why, How, How Much, What, Where, and When (Bertolino, 2007). Why are we looking for errors or trying to improve the usability of the user interface? How do we select the test samples? How (Much) big should the sample be that we will use for testing? What type of test are we going to perform: a unit test, a component/subsystem test, or an integration test? Where are we going to perform the test, in the simulated environment or in the final context? When, at which stage of the product life cycle, will we perform the testing?

Testing does not always get the appropriate emphasis due to time and/or cost constraints (Holzworth et al., 2015). However, a test plan always should be part of integrated model development. The test plan should outline clearly the scope of testing and incorporate test cases, the list of functionalities to be tested, and by whom and when the testing will be performed. Depending on the context, testing can be done by developers, software testers, or end users. Testing of integration can be done with bottom-up or top-down approaches (Myers et al., 2011). Bottom-up testing begins with testing system components at the lowest level and progresses to higher-level combinations of components. In top-down testing, testing begins with the highest-level components and continues to individual components at lower levels. Once the testing strategy is selected, testing procedures can be automated, which has advantages of saving time and assuring repeatability. The challenge is that it may not be possible to automate everything we would like to test, and automated testing cannot always replace manual testing (Berner et al., 2005).

13.3.6 Discoverability, reusability, and interoperability

Obviously, the discoverability and accessibility of stand-alone models is a significant aspect of the model integration process. Despite the large number of modeling components, potential users generally lack the tools to locate them. And if discovered the next challenge is to gain operational access and evaluate their relevance to the user's problem (Booth et al., 2011). Finally, ease of use is important when deciding about model incorporation into a user's integrated system. Preparing software for easy use requires knowledge of the needs of target user groups of the framework.

With the advances of web technology, presenting model interfaces and integration frameworks in web pages has become more popular since it significantly increases model discoverability, accessibility, and usability. Web-based user interfaces that target specific communities, for example, scientists, policy makers, or the public, improve access to data and models and facilitate model reuse. However, providing model via the web is not without challenges. These challenges include managing large volumes of data, complexity of the required software platform and computational demand for model execution, all of which are barriers that limit sharing and reuse of model over the web (Brooking and Hunter, 2013). Developing a flexible architecture to link distributed components, minimizing interoperability agreements, optimizing performance, and securing availability in the long term also are challenges of presenting models on the web (Nativi et al., 2013).

Some modeling frameworks use purposely designed semantic technology features to improve discoverability of models and related resources. During simulation, such systems automatically track input datasets, tools used, parameter values, output datasets, and a series of analysis steps that it stores as history. The user can add annotation about analysis steps, which supports reproducibility of experiments. Incorporation of purposely designed semantic technology features can improve discoverability of model beyond that provided by simple search operations.

13.4 Potential dangers of model integration

Obviously, the idea of integrating existing models has great merit, and the practice of model integration is virtually mandated by the complexity and urgency of the ecological and environmental problems with face. The strategy of leveraging existing stand-alone models by making them discoverable, reusable, and interoperable ideally facilitates holistic, multidisciplinary approaches to these problems. But there are potential dangers, both from theoretical and practical points of view, lurking behind the sophisticated technological advances in computing hardware and software that facilitate model integration. We hasten to add that these dangers are not unique to integrated modeling, they are universal dangers inherent in the modeling process. But as the complexity of the models that are to be integrated, and the sophistication of the process by which they are to be integrated, increases, these dangers become more difficult to recognize and more easily overlooked. Below, we first offer a brief reflection on what we consider is the most insidious potential danger of model integration from a philosophical point of view. Then, we reflect upon what we consider is the most problematic potential danger of integrating models from a practical point of view.

Theoretically, the most insidious potential danger is that of disregarding the implications of the general systems property of Holism, which is among the most basic properties comprising a General Systems Theory of open systems (Von Bertalanffy, 1955; Litterer, 1969, cited in Skyttner, 2005). Simply stated, Holism states that (1) the whole is more than the sum of the parts, (2) the whole defines the nature of the parts, (3) the parts cannot be understood by studying the whole, (4) the parts are interrelated dynamically, and (5) the whole cannot by understood by studying the parts separately. Based on these five statements, it is theoretically impossible to integrate previously built, stand-alone models in a manner such that we can understand them. That is, if we replace "whole" with "integrated model" and "parts" with "previously built, stand-alone models," not only will we not be able to understand the integrated model, we no longer will be able to understand the previously built, stand-alone models due to the dynamic interrelationships among them. Of course, by definition, the General Systems Theory of open systems applies to any open system, and, hence, to any model of an open system, whether integrated or integral. In the very first section of this book, we stated that an important fact to keep in mind is that it is impossible theoretically to predict the future (with a model or otherwise) with certainty (Chapter 1, Section 1.1). Our intent was not to suggest that we should not use models to try to predict the future states of systems. But, rather, we wanted to emphasize the importance of always keeping that fact in mind as we develop, evaluate, and apply simulation models to try to predict the future. Likewise, here, our intent is not to suggest that we should not assemble models from arbitrarily selected parts of a system, we have no choice but to do so. Rather, we want to emphasize the importance of always keeping in mind the necessarily arbitrary nature of our selection process, and that the appropriateness of each part we include in our model, and its connections to other parts of the model, should be judged with reference to the specific objectives of our model. Otherwise, we risk inadvertently including unknown, inappropriate parts and/or inadvertently establishing unknown, inappropriate connections.

From a more practical point of view, the most problematic potential danger of integrating stand-alone models takes the form of solving the technical problem of achieving software compatibility but not achieving the appropriate integration of the contextual and conceptual bases of the original stand-alone models. That is, we might, in fact, inadvertently include unknown, inappropriate parts and/or inadvertently establish unknown, inappropriate connections in our integrated model. This potential danger has been elucidated well by Voinov and Shugart (2013), who coined the term "integronsters" (integrated monsters) and, in addition to word descriptions, provided images of artwork that clearly illustrate the primary danger. Although the pieces can be made to fit together, they do not necessarily make a pleasing (useful) whole. For example, a mermaid is half woman and half fish. A mermaid represented as having the top half of a woman and the bottom half of a fish is (for most people) a pleasing image. However, a mermaid represented as having the top half of a fish and the bottom half of a woman is (for most people) grotesque (see Fig. 1 in Voinov and Shugart, 2013). Likewise, treating models only as software when solving the integration problem can result in constructs that are perfectly valid as software products but useless, or, worse, misleading, as models. Solving the technical problem of software compatibility does not guarantee appropriate integration of the scientific knowledge bases represented in the original stand-alone models. That is, resolving problems such as those related to conversion from different programming languages, poor code documentation, different dimensionalities, and asynchronous time stepping, although difficult, is not as challenging as achieving the appropriate integration of the contextual and conceptual bases of the original stand-alone models. The context of a model is the domain over which the model is intended to operate, which embodies all of the assumptions under which the model operates, and refers especially to those assumptions that remain implicit (see Chapter 10, Section 10.7.1). The conceptual basis of a model depends directly upon the purpose of the model (the model objective) and the model domain. The spatial and temporal scales of the model, as well as the levels of detail with which model components are represented, both define and justify the conceptual basis of the model. As we emphasized in Chapter 6 and reflected upon in Chapter 10, our statement of purpose provides the framework for model development, indicates the standards for model evaluation, and suggests the framework within which simulation results should be interpreted. Thus, any shift in context or purpose, such as might occur as a result of integration with other models, would cast doubt on the relevance of previous model evaluation procedures.

From both theoretical and practical points of view, the potential dangers of integrating individual parts into a model are unavoidable. Theoretically, the potential dangers are identical for integrated and integral models. In practice, we would suggest that the danger level increases as the complexity of the parts that are to be integrated, and the sophistication of the process by which they are to be integrated, increases. The basic danger was identified before the integration of large stand-alone models was technically feasibility (before we had the necessary hardware and software). For example, Ågren and Bosatta (1990) put forth an argument related to empirically based quantitative ecological modeling in general that encapsulated the basic problem. They stated that the complexity of ecological systems makes it unlikely that we can understand system properties by fitting system parts together empirically piece by piece, and that such properties are more likely to be understood via direct investigation of the system as a whole. They offered a simple numerical example to demonstrate that it is not trivial to synthesize a larger system from its component parts and infer the system properties from the properties of the parts, even when the parts themselves are trivially simple. Their example system consisted of two components, $X1$ and $X2$. Each component is drained by two flows following first-order kinetics. If component 1 is operating separately, it is described by the equation:

$$\mathrm{d}X_1/\mathrm{d}t = -k_1X_1 - k_{21}X_1 = -(k_1 + k_{21})X_1$$

and similarly for X_2. The time constant for component 1 is then: $(k_1 + k_{21})^{-1}$, and for component 2, $(k_2 + k_{12})^{-1}$. They assumed for simplicity that $k_1 + k_{21} = k_2 + k_{12} = k$. When the two components operate as a system, one of the flows from each of the components is directed to the other component, giving for component 1 the equation:

$$\mathrm{d}X_1/\mathrm{d}t = -k_1X_1 - k_{21}X_1 + k_{12}X_2$$

and similarly for component 2. It is now easy to show that the time constants for the system are $[k \pm (k_{21}k_{12})^{1/2}]^{-1}$. These time constants can be very different from those of the constituent parts. Thus, our way of looking at the whole-system behavior can be entirely different from the way we look at the parts. In this example, the whole system will respond to different frequencies in the environment than will the parts. Ågren and Bosatta (1990) contended that if we wanted to understand an ecological system as such, and derive properties that are valid at the system level, we might be better off starting directly at the whole-system level, forgetting about the properties of the individual parts in isolation. These authors argued for more artistry and less construction engineering in ecology, evoking an image that perhaps in a way foreshadowed Voinov and Shugart's (2013) concern about "integronsters." They also noted that the increasing capacity of computers and simulation programs soon would provide modeling capability on every scientist's desk, and expressed concern that computer technology might actually hinder scientific thought. That concern merits our continued reflection.

13.5 Future of integrated models

The rationale for integrating existing models within the context of ecological and environmental modeling is that we need to integrate information from a variety of scientific fields to address the complex, multidisciplinary problems that confront us in today's world. The rationale is compelling, the use of integrated models is commonplace in today's environmental management programs, and there is no reason to believe that use will diminish in the future. The most insidious potential danger of integrating stand-alone models results from solving the technical problem of achieving software compatibility but not achieving the appropriate integration of the diverse scientific bases of the original stand-alone models. This danger is not unique to integrated modeling, it is a universal danger inherent in the modeling process whenever information from diverse scientific fields is included in a model. However, we would suggest that there is a difference between "integrated" and "integral" modeling, as we defined these terms in Section 13.1, with regard to the level of danger, or the inherent risk, involved. The difference in danger levels is due to the difference in the manner in which the communication among experts/modelers from the different scientific disciplines takes place. During integral modeling, most commonly the communication among the modelers responsible for coding the different scientific components of the model is continuous from the beginning. Problems arising from contextual and semantic differences among scientific disciplines can be resolved via such face-to-face communication (often among a relatively small number of people). During integrated modeling, most commonly there will have been little or no communication among the teams of modelers that developed the various original stand-alone models, and there will be little or no possibility of getting all the members of these teams together to communicate face-to-face (or at least "live" via the web). Ideally, these contextual and semantic problems are resolved during the preintegration assessment phase of the integrated modeling process (see Section 13.3.1). However, to date, as noted by Belete et al. (2017) in their recent review, the preintegration phase of most integrated modeling projects has not been described well.

That the biggest current challenge facing integrated modeling is related to the preintegration assessment phase is widely recognized. Belete et al. (2017) concluded that the connection between software integration frameworks and science domains cannot and should not be ignored. Of the 12 sets of questions identified by these authors, the answers to which facilitate the pre-integration assessment (see Section 13.3.1), the first four are domain-specific questions. The fourth question specifically asks if the required modeling components are from the same science domain, if the semantics are compatible, and if available software frameworks cover the relevant science domains or a new framework must be developed.

Based on their extensive review of the recent literature, Belete et al. (2017) offered several recommendations for moving forward as a global community in developing integrated modeling systems. Their recommendations reflect the software engineering concept of "separation of concerns" and the science-based concept of ontologies. Separation of concerns focuses on developing software components that perform a specific limited set of functionalities. As more functionality is packaged in a single piece of software, its flexibility and use (and reuse) becomes more and more constrained in an integrated context. Ontologies are a structured way to represent knowledge about a domain. Ontologies define and describe concepts (things) included within a knowledge domain, as well as their related properties and relationships. Belete et al. (2017) identified three principal elements of integrated modeling systems: a user interface, a model integration infrastructure and tools (i.e., a software framework), and a set of science-based components.

The user interface should be "thin" and focus on specific problems or classes of problems that give users the ability to select or construct a scientific workflow involving data, modeling, and tool components relevant to the user-specific problem, and visualize and process results of workflow execution. Workflows should accommodate at least two levels of complexity. First, a straightforward linking of multiple data, model, and tool components to enable both feed forward and feedback interoperability. Second, it should accommodate special system-level functionality such as data acquisition and harmonization tools, ability to perform system-wide sensitivity and uncertainty analysis, Monte Carlo simulation, and analytical tools for synthesizing intermediate and endpoint data processing. These workflows define the requirements for design of the model integration framework and its components. Users should have access to a repository of components and sufficient metadata about them to determine which ones are relevant to a user-stated problem and are interoperable. Metadata about model components (meta-models) should be standardized to ensure machine-readability and processing. The repository may be local (to an organization or a development group, for example) but, ideally, would be linked to a global collection of socioenvironmental science software components.

Output of the user interface should be a stand-alone (and, ideally, standards-based) description of the workflow to be executed. The interface should also provide access to tools that consume and process results of the integrated modeling execution.

The model integration framework implements workflow produced via the user interface. Important elements of the integration process include execution management, semantic mediation, dataset conversion, and error handling. Integration tools or complete frameworks may be designed to optimize model system execution on high performance computers, parallelization of execution (e.g., leveraging the cloud), and web-based integration.

The set of science components should be designed and implemented in a framework-independent manner. Individual scientists should be encouraged to publish software in a way that can be discovered, understood, and integrated within any user interface and model integration system. Essential features associated with the design of components to meet these high-level requirements include the following. (1) An explicit API (application program interface). At the highest level, the API enables initializing, running, and closing functions. At a secondary level, it describes and enables programmatic access to science knowledge and functionality expressed in the component. (2) Metadata for input/output/internal variables that describe at least the meaning of variables, including assumptions made about variables and processes; unambiguous units of measure; and the valid range of values, if appropriate. (3) Within the context of (1) and (2), the content should be described using a controlled vocabulary and, eventually, a full ontology to describe relevant concepts and relationships. (4) Open-source software available via public access version control systems such as gitHUB. To develop a global repository, components must be published with science content and functionality to enable discovery, understanding, and integration. Scientific publications must include metadata that fully describe inputs, outputs, and internal workings of the component. This will require controlled vocabularies and an ontological framework for describing data and modeling science. To achieve this, the modeling community should reinforce collaboration of developing and improving standards that can be used across software frameworks.

Alternative ecological modeling formats: Commonalities from a systems perspective

Chapter outline

14.1 Overview

In this chapter, we reflect upon the commonalities among several alternative quantitative ecological modeling formats from a systems perspective. These include matrix models, network models, and individual-/agent-based models. At first glance, these alternative formats differ markedly from the compartment model format we have been using throughout this book. However, we would suggest that all are

Developments in Environmental Modeling, Volume 31, ISSN 0167-8892, https://doi.org/10.1016/B978-0-444-64163-2.00014-1

fundamentally similar when viewed from the systems perspective. That is, although details of the calculations involved in executing these types of models do, indeed, differ, all are based on the two most fundamental characteristics of systems: (1) the whole is more than the sum of the parts, and (2) system structure is the source of system behavior (see Chapter 1, Section 1.4). In the following sections, we briefly describe each format, and then offer a synthesis of their commonalities from a general systems perspective. Note that our intent is not to examine an exhaustive list of possible modeling formats, nor to examine in detail the more sophisticated quantitative procedures supported by the particular formats just mentioned. Our intent is simply to emphasize that all of these formats, as well as other quantitative formats for representing the dynamics of complex systems, fit comfortably within the systems approach to problem-solving and do not alter the general methodology for developing, evaluating, and applying ecological models we have described in this book.

14.2 Matrix models
14.2.1 General background

In mathematics, a matrix is a rectangular array of elements (for our purposes, numbers) arranged in rows and columns. For example, a 3×2 matrix has 3 rows and 2 columns, a 2×2 matrix has 2 rows and 2 columns, and a 3×1 matrix has 3 rows and 1 column. The latter is referred to as a column vector (a 1×3 matrix is referred to as a row vector). The specific positioning of the elements relative to one another is important. There is a variety of simple mathematical procedures associated with matrix algebra, such as matrix multiplication, as well as some more complicated procedures, such as the calculation of eigenvalues and eigenvectors, which have been used in ecological matrix models to represent how the system-of-interest changes over time and to indicate the relative influence of different model parameters on the rate of change of the system (Caswell, 2006). The most basic procedure employed is to premultiply a square matrix times a column vector (premultiply means that the matrix is on the left, or comes before, the vector), which results in another column vector, as illustrated in Fig. 14.1A. Note that the number of columns in the matrix must equal the number of rows in the vector it multiplies. Also note that the number of rows in the resulting vector equals the number of rows in the original vector. Thus, we could premultiply the matrix times the vector resulting from the first multiplication and generate another new vector. In fact, we could repeat this procedure indefinitely. Or in other, more technical, words, we could repeatedly premultiply the transition matrix times the current vector generating a Markov Chain of vectors in which the values in each new vector depend only on the values in the transition matrix and the values in the current vector (Fig. 14.1B).

A
$$\begin{bmatrix}1 & 2 & 3\\4 & 0 & 0\\0 & 5 & 0\end{bmatrix}\begin{bmatrix}3\\2\\1\end{bmatrix}=\begin{bmatrix}1\times3+2\times2+3\times1\\4\times3+0\times2+0\times1\\0\times3+5\times2+0\times1\end{bmatrix}=\begin{bmatrix}10\\12\\10\end{bmatrix}$$
$$M \qquad v_0 \qquad\qquad\qquad\qquad\qquad v_1$$

B
$$Mv_0 = v_1$$
$$Mv_1 = v_2$$
$$\vdots \qquad \vdots$$
$$Mv_n = v_{n+1}$$

FIG. 14.1

Illustration of (A) the procedure employed to premultiply a square matrix times a column vector, which results in another column vector, and (B) the repeated premultiplication of a transition matrix times the current vector creating a Markov Chain of vectors in which the values in each new vector depend only on the values in the transition matrix and the values in the current vector.

14.2.2 Ecological applications

In ecological matrix models, we might think of the elements in the transition matrix as representing the rules governing the change of the state of the system, and the elements in the vector as representing the current state of the system. Thus, the change in the state of the system depends on the current state of the system. Among the most common ecological applications of matrix models is to study the dynamics of age-structured populations using some adaptation of the Leslie matrix (Leslie, 1945). The Leslie matrix is a square transition matrix that contains age-specific natality rates (F_x) as elements of the first row, and age-specific survival rates (P_x) as subdiagonal elements. F_x usually is defined as the number of female offspring born per female aged x that survive to enter age 0 at the beginning of the next time interval, and P_x is defined as the proportion of individuals aged x that survive to enter age $x+1$ at the beginning of the next time interval. Premultiplication of the Leslie matrix times a column vector that contains as its elements the number of individuals in each age class within the population at time t results in a column vector that contains as its elements the number of individuals in each age class within the population at time $t+1$. Premultiplication of the Leslie matrix times successive age-class vectors generates a Markov Chain of vectors which represents a projection of the age structure and size of the population over time (Fig. 14.2). Since the elements in the Leslie matrix are constants, the simulated population will attain its stable age distribution (meaning the proportion of the total number of individuals in each age class will remain constant) and will either increase exponentially, decrease exponentially, or remain exactly constant. The stable age-class distribution and the finite rate (λ) of population growth (or decline) also can be derived analytically via analysis of the Leslie matrix, as can various other matrix attributes of ecological interest, such as the sensitivity of λ to changes in the values of the matrix elements representing the age-specific natality and survival rates (Caswell, 2006).

$$\begin{bmatrix} f_0 & f_1 & f_3 \\ s_0 & 0 & 0 \\ 0 & s_1 & 0 \end{bmatrix} \begin{bmatrix} A_0 \\ A_1 \\ A_2 \end{bmatrix}_t = \begin{bmatrix} A_0 \\ A_1 \\ A_2 \end{bmatrix}_{t+1}$$

$$\sum A_i = N$$

Stable age-class distribution: $\begin{bmatrix} A_0/N \\ A_1/N \\ A_2/N \end{bmatrix}_{t+n} = \begin{bmatrix} A_0/N \\ A_1/N \\ A_2/N \end{bmatrix}_{t+n+1}$

FIG. 14.2

Illustration of the premultiplication of a Leslie matrix times successive age-class vectors to generate a Markov Chain of vectors which represents a projection of the age structure and size of a population over time.

There are a variety of other ecological applications of matrix models. For example, models of landscape change using discrete state spaces can be expressed in their simplest form as a matrix model as: $v_{t+1} = Mv_t$, where v_t is a $n \times 1$ column vector whose elements represent the fraction of land area in each of n states at time t, and M is an $n \times n$ transition matrix whose elements represent rates of change of each state during the time interval from t to $t+1$ (Baker, 1989). In general, any "state and transition" type of model could be represented as a matrix model (Caswell, 2006; Leslie, 1945; Lefkovitch, 1965; Baker, 1989).

14.3 Network models

14.3.1 General background

A network is any collection of units potentially interacting as a system (Proulx et al., 2005). The simplest type of network is composed of a set of uniform nodes connected by undirected edges (Fig. 14.3A). But there are not necessarily edges between each pair of nodes (Fig. 14.3B), and edges can be directed, indicating a one-way effect, or a one-way flow, between two connected nodes (Fig. 14.3C). Thus, attributes of a network clearly depend on its structure. For example, the shortest path from Node A to Node C in Fig. 14.3A, B, and C is of length 1, 2, and 2, respectively. Nodes can represent virtually any entity and edges can represent virtually any interaction between pairs of nodes. Within an ecological modeling context, nodes might represent individuals in a population or species in a community, and edges might represent interactions between pairs of individuals or between pairs of species. There is a variety of mathematical procedures that can be used to summarize quantitatively the properties and behavior of a network via a few simple statistics. These include average path length/diameter (the mean shortest path between all nodes in the network), betweenness (the number of shortest paths that the focal node lies on), closeness (the mean shortest path between a focal node and all other nodes in the network), clustering

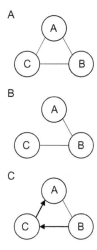

FIG. 14.3

(A) The simplest type of network represented by a set of uniform nodes connected by undirected edges. (B) A simple network represented by a set of nodes in which not all of the nodes share the same number of connections. (C) A slightly more complex network represented by a set of nodes connected by both undirected edges and directed edges. *Thick lines* represent the shortest path between Node A and Node C.

coefficient (the proportion of a focal node's neighbors who are themselves neighbors), degree/connectivity (the number of edges that connect the focal node to other nodes), and degree distribution (the frequency distribution of the individual node degree for an entire network) (Proulx et al., 2005).

14.3.2 **Ecological applications**

We might describe ecological network models as a systems-oriented methodology to analyze within-system interactions that define holistic system properties, such as points of accumulation of energy and pathways of energy flow through a food web (Fath et al., 2007). Among the most common ecological applications of network models is to study the relationship between food web structure and community stability (see review by Proulx et al., 2005). Ecologists have used dynamic network models to explore how size and connectivity of food webs affect stability and long-term persistence of communities in the face of fluctuations in density of constituent species and/or invasion of new species, and, in turn, how removing or replacing native species with exotic invaders can alter food-web structure. The effects of these perturbations on community stability (on network robustness) can be measured in terms of the number of secondary extinctions they cause and

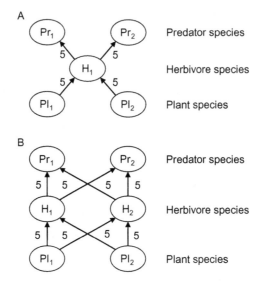

FIG. 14.4

Illustration of network models representing systems in which (A) a single herbivore species acts as an energy bottleneck between plants and predators, and (B) such a bottleneck does not exist due to the presence of more than one herbivore species. Assumptions for both models are that each herbivore species can obtain a maximum of 10 units of energy per unit time from plants and that each predator species can obtain a maximum of 10 units of energy per unit time from herbivores.

by the break-up of the network into smaller components (Proulx et al., 2005). Conceptualizing food webs as energy-flow networks allows their description not only in terms of direct trophic interactions but also in terms of indirect effects caused by "dominating" nodes that act as energy bottlenecks. For example, a single herbivore species might act as an energy bottleneck between plants and predators. Generalizing this idea, we might expect highly connected species in a food web to cause more secondary extinctions if they are removed than sparsely connected species (Fig. 14.4).

There are a variety of other ecological applications of network models, notably, in the field of theoretical systems ecology. Within this context, ecological network analysis provides a system-oriented perspective to identify and quantify direct and indirect relationships among all of the objects in a system (Patten, 1985). This work has quantified several ecological network properties that have been used to investigate ecosystem behavior, including amplification, indirect effects, homogenization, and synergism, which together describe the distribution of energy-matter flow through the network's many direct and indirect pathways (Fath and Patten, 1999). Most commonly, these properties are identified under steady-state conditions, that is, after the system has reached equilibrium. An oversimplified summary of the main findings of ecological network analysis is that ecosystems try to capture as much energy as they can, and retain it (recycle the captured energy among ecosystem components) as long as they can.

14.4 Individual-/agent-based models

14.4.1 General background

Individual-based models (IBMs) and agent-based models (ABMs) are simulation models that describe the behavior of autonomous individuals (organisms) or agents, respectively (Grimm et al., 2006). These models have become widely used, not only in ecology but also in other disciplines dealing with complex systems made up of autonomous entities (see Grimm et al., 2006 for references). ABM is a more widespread and generic term for IBM which is used most commonly in fields other than ecology. The autonomous entities in ABMs do not necessarily need to be individual organisms, they may social groups within a community, corporations within an industry, or any other entity that might be defined as operating autonomously among other autonomously operating entities within the system-of-interest. IBMs, for the purposes of this chapter, we define following Grimm et al. (2006). An IBM is a model of a system of individuals and their environment in which system behavior arises from traits of the individuals and characteristics of the environment. Thus IBMs allow researchers to study how system-level properties emerge from the behavior of individuals, as well as how the current state of the system affects the behavior of individuals.

14.4.2 Ecological applications

IBMs can represent explicitly the variability among individuals, the local interactions among individuals, the interactions between individuals and their local environment, and the effect of internal physiological conditions on the behavior of individuals (Fig. 14.5). The manner in which emergent system-level properties arising from individual behaviors is summarized depends on model objectives, but might include such system-level variables as population size, sex ratio, age structure, ratio of infected to uninfected individuals, predator/prey ratio, species richness, and/or species diversity, to name only a few (Fig. 14.5). The wide array of ecological questions that have been addressed using IBMs includes those related to group and social behavior, population dynamics of social animals, dispersal and habitat selection, regulation of hypothetical populations, dynamics of plant populations and communities, and structure of communities and ecosystems (see Table 6.1 in Grimm et al. 2006 for references).

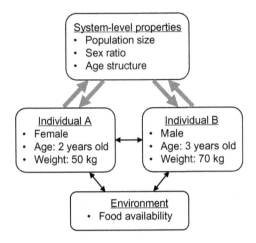

FIG. 14.5

In an individual-based model (IBM), system-level properties emerge from the interactions among individuals and with their environment, and the current state of the system (embodied in the system-level-properties) affects the behavior of individuals.

14.5 Commonalities among formats from a systems perspective

14.5.1 Basic characteristics of systems common to systems models

In Chapter 1 (Section 1.4), we stated that all systems, ecological or otherwise, have five important characteristics in common: (1) A system is more than the sum of its parts. (2) System structure is the source of system behavior. (3) Systems are hierarchical. (4) Systems are resilient. (5) Systems are self-organizing. We also stated that not all systems models exhibit each of these five system characteristics, but, by definition, all exhibit the first two. Above, we noted that a matrix is a rectangular array of elements and that result of matrix multiplication depends on the positioning of the elements in the matrix (Section 14.2), that a network is composed of a set of nodes connected by edges and that the attributes of a network depend on the structure of these connections (Section 14.3), and that an IBM is composed of autonomous individuals and their environment and that system behavior arises from interactions among individuals and with their environment (Section 14.4). Thus, although details of the calculations involved in executing these alternative types of models differ, for each, as for compartment models, system structure is the source of system behavior, and system behavior cannot be understood by examining the behavior of individual system components in isolation from the rest of the system (we provide some simple numerical examples using each of these four formats in Section 14.5.3).

From the systems perspective, regardless of the quantitative format chosen, the general principles and guidelines for developing ecological models remain the same. "Modeling ecological systems" does not connote use of a specific quantitative methodology. "Modeling ecological systems" should connote use of the systems perspective, or systems approach, which embodies both a philosophical perspective and a *collection* of quantitative techniques that emphasize a holistic approach to solving problems involving complex systems. That is, we should not become so fascinated with quantitative methodological details that we lose our systems perspective on the problem. But, rather, we should identify the most appropriate quantitative modeling format from our systems perspective on problem at hand.

14.5.2 **Commonality does not connote interchangeability**

We hasten to note that our intention is not to suggest that alternative ecological modeling formats are completely interchangeable. For any particular modeling application some formats will be more appropriate than others (Table 14.1). Matrix models and network models are amenable to analytical mathematical solutions under steady-state conditions that can have useful ecological interpretations, some of which we mentioned in Sections 14.2.2 and 14.3.2, respectively. Analytical mathematical solutions generally are not possible for compartment models and are virtually impossible for IBMs. These models must be solved numerically via simulation. On the other hand, matrix models and network models do not possess the flexibility provided by compartment models and IBMs to represent the details of ecological processes explicitly. Also, the explicit representation of spatial heterogeneity and individual variability is awkward in matrix models and network models (requires sophisticated mathematics), and cumbersome in compartment models ("bookkeeping" associated with maintaining multivariate descriptions of individuals becomes computationally intense). Representation of stochasticity is awkward in matrix models and network models (requires sophisticated mathematics), but easy in compartment models and IBMs. Of course, there are numerous exceptions to these broad generalizations. Nonetheless, we believe the general attributes and typical applications provided in Table 14.1 are indicative of the compartment models, matrix models, network models, and IBMs most commonly encountered within an ecological context.

14.5.3 **Numerical examples of commonalities among formats**

In this section, we provide simple numerical examples demonstrating the commonalities among compartment models, matrix models, network models, and IBMs from a systems perspective. That is, we provide numerical examples demonstrating that (1) system structure is the source of system behavior, and (2) system behavior cannot be understood by examining the behavior of individual system components in isolation from the rest of the system, regardless of the specific format used.

Table 14.1 Several general attributes, a typical ecological application, and a general reference for compartment models, matrix models, network models, and individual-based models.

Format	Analytical solutions common?	Temporal variability easily represented?	Spatial heterogeneity easily represented?	Individual variability easily represented?	Stochasticity easily represented?	Typical ecological application	General reference
Compartment models	No	Yes	No	No	Yes	Many	This text
Matrix models	Yes, under steady-state conditions	Yes	No	No	No	Dynamics of age-/stage-structured populations	Caswell (2006)
Network models	Yes, under steady-state conditions	Yes	No	No	No	Energy/biomass flow through food webs	Proulx et al. (2005)
Individual-based models	No	Yes	Yes	Yes	Yes	Many	Grimm and Railsback (2005)

Consider the following system-of-interest. A hypothetical population of 100 individuals is exposed to a disease-causing pathogen. Initially, none of the individuals carry the pathogen, but all are susceptible (S) to becoming infected (I). All infected individuals eventually will recover (R), but recovered individuals again become susceptible to the disease. (You may recognize this as the classical SIRS model frequently used in epidemiology.) We are interested in simulating the dynamics of this disease within the population. In particular, we are interested in how the proportion of the population in each of these three states (S, I, and R) varies over time. We know the daily infection rate (the rate at which S individuals are transformed into I individuals, $t_{si} = 0.625$), the daily recovery rate (the rate at which I individuals are transformed into R individuals, $t_{ir} = 0.6$), and the daily rate at which immunity to the disease is lost (the rate at which R individuals are transformed into S individuals, $t_{rs} = 0.1$).

Compartment model

Using a compartment model format (Fig. 14.6A), we could write the following equations to represent the dynamics of the system-of-interest:

$$S_{t+1} = S_t + (rs - si)\,\Delta t \tag{14.1}$$

$$I_{t+1} = I_t + (si - ir)\,\Delta t \tag{14.2}$$

$$R_{t+1} = R_t + (ir - rs)\,\Delta t \tag{14.3}$$

$$si = t_{si} \times S \tag{14.4}$$

$$ir = t_{ir} \times I \tag{14.5}$$

$$rs = t_{rs} \times R \tag{14.6}$$

where S, I, and R represent state variables, si, ir, and rs are material transfers, and t_{si}, t_{ir}, and t_{rs} are constants. Using these equations to simulate disease dynamics within the population, we would obtain the temporal variation of the proportions of S, I, and R individuals in the population shown in Fig. 14.6B.

Matrix model

Using a matrix model format, we could write the equations to represent the dynamics of the system-of-interest as shown in Fig. 14.7A, where S, I, and R are elements of the state vector, and t_{si}, t_{ir}, and t_{rs}, as well as $(1 - t_{si})$, $(1 - t_{ir})$, and $(1 - t_{rs})$ are elements of the transition matrix. Using these equations to simulate disease dynamics within the population (Fig. 14.7B), we would obtain exactly the temporal variation of the proportions of S, I, and R individuals in the population shown for the compartment model in Fig. 14.6B.

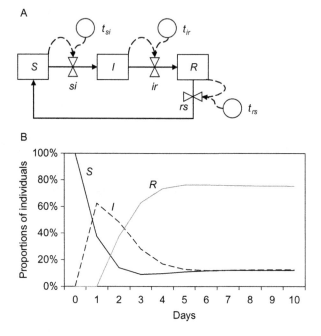

FIG. 14.6

(A) Compartment model representing disease dynamics, where S represents the number of susceptible individuals, I represents the number of infected individuals, and R represents the number of recovered individuals; t_{si}, t_{ir}, and t_{rs} represent daily rates of infection, recovery, and loss of immunity, respectively; and si, ir, and rs represent daily numbers of individuals transformed from S to I, from I to R, and from R to S, respectively. (B) Temporal variation of the proportions of S, I, and R individuals in the population.

A

$$
\begin{bmatrix} 1-t_{si} & 0 & t_{rs} \\ t_{si} & 1-t_{ri} & 0 \\ 0 & t_{ri} & 1-t_s \end{bmatrix} \begin{bmatrix} S_t \\ I_t \\ R_t \end{bmatrix} = \begin{bmatrix} S_{t+1} \\ I_{t+1} \\ R_{t+1} \end{bmatrix}
$$
$$
\quad\quad M \quad\quad\quad\quad v_t \quad\quad v_{t+1}
$$

B

$$
\begin{bmatrix} 0.375 & 0 & 0.1 \\ 0.625 & 0.4 & 0 \\ 0 & 0.6 & 0.9 \end{bmatrix} \begin{bmatrix} 100 \\ 0 \\ 0 \end{bmatrix}_t = \begin{bmatrix} 37.5 \\ 62.5 \\ 0 \end{bmatrix}_{t+1}
$$

$$
\begin{bmatrix} 0.375 & 0 & 0.1 \\ 0.625 & 0.4 & 0 \\ 0 & 0.6 & 0.9 \end{bmatrix} \begin{bmatrix} 37.5 \\ 62.5 \\ 0 \end{bmatrix}_{t+1} = \begin{bmatrix} 14.06 \\ 48.44 \\ 37.50 \end{bmatrix}_{t+2}
$$

FIG. 14.7

(A) Matrix model representing disease dynamics, where S represents the number of susceptible individuals, I represents the number of infected individuals, and R represents the number of recovered individuals; and t_{si}, t_{ir}, and t_{rs} represent daily rates of infection, recovery, and loss of immunity, respectively. (B) Illustration of the matrix calculations yielding the temporal variation of the proportions of S, I, and R individuals in the population. Only the first two of a series of iterative calculations are shown.

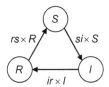

FIG. 14.8

Network model representing disease dynamics, where S, I, and R are nodes representing the numbers susceptible, infected, and recovered individuals, respectively, and edges connecting the nodes, represented quantitatively by $si \times S$, $ir \times I$ and $rs \times R$, represent the daily number of individuals changing disease status from S to I, from I to R, and from R to S, respectively.

Network model

Using a network model format, we might sketch the conceptual model diagram shown in Fig. 14.8, where S, I, and R are nodes, and edges connecting the nodes are represented quantitatively by $si \times S$, $ir \times I$, and $rs \times R$. The calculations to simulate disease dynamics within the population are the same as when using a matrix model format, and we would obtain exactly the temporal variation of the proportions of S, I, and R individuals in the population shown for the compartment model (and the matrix model) in Fig. 14.6B.

Individual-based model

Using an IBM format, we might sketch the conceptual model diagram shown in Fig. 14.9, where S, I, and R represent the current disease status of an individual as being either susceptible, infected, or recovered, respectively, and t_{si}, t_{ir}, and t_{rs} represent environmental characteristics that determine the daily probabilities that individuals will become infected, recover, and lose immunity, respectively. That is, daily probabilities that the disease status of an S individual will change to I, that the disease status of an I individual will change to R, and that the disease status of an R individual will change to S, respectively. We need to write computer code to execute the model, to represent the dynamics of the system-of-interest. The syntax of the code would differ depending on the software we were using. For example, if we were using NetLogo (https://ccl.northwestern.edu/netlogo/), we might write code similar to that in Fig. 14.10. The code is not understandable, of course, unless you "speak" NetLogo. We offer it only as a tangible example to demystify what an IBM looks like. You may recognize some "S I Rs" sprinkled through the code, as well as the numerical values of the infection, recovery, and loss of immunity rates. In any event, execution of this, or equivalent, code will yield temporal variation of the proportions of S, I, and R individuals in the population quite similar to that shown for the compartment (and matrix and network) model in Fig. 14.6B. The IBM simulation results will differ slightly because discrete individuals (not fractions thereof) are infected, recover, and loss their immunity, and also because the model is stochastic. Each day, each individual has a probability of changing disease status. In the compartment,

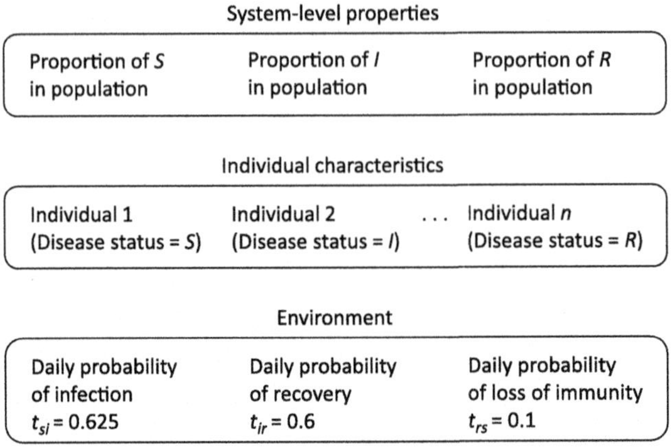

System-level properties

Proportion of S in population	Proportion of I in population	Proportion of R in population

Individual characteristics

Individual 1 (Disease status = S)	Individual 2 (Disease status = I)	...	Individual n (Disease status = R)

Environment

Daily probability of infection $t_{si} = 0.625$	Daily probability of recovery $t_{ir} = 0.6$	Daily probability of loss of immunity $t_{rs} = 0.1$

FIG. 14.9

IBM representing disease dynamics, where S, I, and R represent the current disease status of an individual as being either susceptible, infected, or recovered, respectively, and t_{si}, t_{ir}, and t_{rs} represent environmental characteristics that determine the daily probabilities that individuals will become infected, recover, and lose immunity, respectively. That is, daily probabilities that the disease status of an S individual will change to I, that the disease status of an I individual will change to R, and that the disease status of an R individual will change to S, respectively.

matrix, and network models, proportions of the numbers of S, I, and R individuals changed disease status each day.

Thus, as we noted at the beginning of this chapter (Section 14.1), system structure is the source of system behavior regardless of the specific quantitative modeling format used. The connections among the state variables in the compartment model, the positioning of the elements in the transition matrix in the matrix model, the connectivity of nodes in the network model, and the traits and behaviors of individuals coded into the IBM determined, in their respective models, the general behavior of the system-of-interest.

We also noted at the beginning of this chapter that system behavior cannot be understood by examining the behavior of individual system components in isolation from the rest of the system, regardless of the specific quantitative modeling format used. By now, this should be obvious, but lest we leave demonstration of such an important point implicit, we conclude this chapter with a brief demonstration based on our SIR models. Consider the dynamics of infected individuals graphed in Fig. 14.11. Each of the two lines can be generated by making the same simple structural change to any of the models (compartment, matrix, network, or individual-based) we examined in this chapter. Why do the two lines differ? What single change to model structure caused this difference in the dynamics of I? Well, this is a very simple model, and it may not have been so difficult to answer the question (and probably trivially easy if you are familiar with SIR models and with a main cause

```
;##############################################################################
globals [t prS PrI PrR]

patches-own
[
  S I R
]

turtles-own [status tsi tir trs already-changed-status]

;##############################################################################
to setup

  clear-all
  reset-timer
  reset-ticks

  ; set initial conditions
  create-turtles 100
  [
    set xcor 1 set ycor 1
    set already-changed-status 0
    set status "S"
    set tsi 0.625
    set tir 0.6
    set trs 0.1
  ]

  ask patch 1 1
  [
    set S count turtles-here with [status = "S"]
    set I count turtles-here with [status = "I"]
    set R count turtles-here with [status = "R"]
    set prS S / (S + I + R)
    set prI I / (S + I + R)
    set prR R / (S + I + R)
  ]

end ; end of "setup"
;##############################################################################
```

FIG. 14.10

Computer code to execute the IBM representing disease dynamics using NetLogo. The syntax of the code would differ depending on the software used to execute the model.

of an infection "burning out"). We eliminated the material transfer (matrix element, network edge, portion of computer code) representing the loss of immunity. Thus, the dynamics of I reflect the transient behavior of the system as all S eventually become (permanently) R, and there no longer are S to become I. One could imagine that, in a more complex (interconnected) system, the dynamics of a single system component in response to a myriad of indirect effects from unmonitored system components would be impossible to understand.

```
;##############################################################
to go

  ; advance 1 time step
  tick
  set t t + 1

  ;check for end of simulation
  if ticks > 100
  [
  print timer
  stop
  ]

  ; update system
  ask turtles
  [
  set already-changed-status 0
  if status = "S" and already-changed-status = 0 [let change-status random-float 1 if change-status <
tsi [set status "I" set already-changed-status 1]]
  if status = "I" and already-changed-status = 0 [let change-status random-float 1 if change-status <
tir [set status "R" set already-changed-status 1]]
  if status = "R" and already-changed-status = 0 [let change-status random-float 1 if change-status <
trs [set status "S" set already-changed-status 1]]
  ]

  ask patch 1 1
  [
  set S count turtles-here with [status = "S"]
  set I count turtles-here with [status = "I"]
  set R count turtles-here with [status = "R"]
  set prS S / (S + I + R)
  set prI I / (S + I + R)
  set prR R / (S + I + R)
  ]

end ; end of "go"
;##############################################################
```

FIG. 14.10—CONT'D

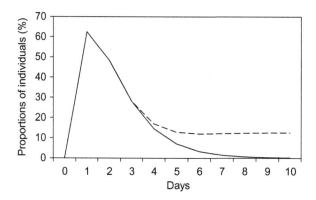

FIG. 14.11

Graph of the proportions of infected individuals (*I*) resulting from two simulations, one run before, and one run after, making a simple change to our SIR model (our compartment, matrix, network, and individual-based models all generate the graphed difference as a result of making the same structural model modification).

What about the art of modeling?

This book, by its title, claims to be an introduction to the art and science of modeling ecological systems. In the preface, we stated that we view ecological modeling as both an art and a science, that the science has its foundations in systems theory, and that the art has its foundations in our innate ability for abstract representation of the world around us. We further stated that we have tried to achieve an appropriate balance between the artistic and scientific aspects of the modeling process. However, arguably, much of the "art" of modeling has remained implicit throughout this book. We would hope that readers have gained a "sense" of the "artistic" aspects of ecological modeling, particularly while reading the chapter on conceptual model formulation (Chapter 4), and also the chapter in which we presented a candid view of the modeling process (Chapter 9). Nonetheless, as the title of this final chapter suggests, we would like to reflect more explicitly on the artistic aspects of modeling, particularly within the context of the science of ecology.

So where do we draw the line between art and science. We might define science as a branch of knowledge or study that deals with the systematic organization of facts showing the operation of general laws (a dictionary definition), or, as systematic knowledge of the physical or material world (also a dictionary definition). We might define art as the principles governing any craft, skill, or branch of learning (a dictionary definition, the fifth alternative definition), or, as the quality, production, expression, or realm of what is beautiful (the first of several alternative dictionary definitions). One can find books titled "The art and science of ..." written about almost any topic you can think of. Without providing a long list of such titles or getting bogged down in semantics, it seems to us that the reason for adding "art" to most of these titles is to convey the idea that there is more to the subject than science can explain, systematically or otherwise.

Decades ago, Ågren and Bosatta (1990) wrote a provocative discussion paper titled "Theory and model or art and technology in ecology." They argued for more artistry and less construction engineering, or more science and less numerical modeling, in ecology. These authors posed the question; if ecology is the science of "the house," how do we as ecologists relate to our house? Are we like construction engineers who mainly care about getting the house built, or like artists who try to capture the essence of the house in our paintings? Both approaches contribute to our use and conception of a house, but they do not contribute in the same manner. Similarly, theory and numerical models provide two approaches to the advancement and application of ecological understanding, but they do not contribute in the same manner. Numerical models (technology) can integrate and apply existing knowledge, whereas theory (art) encompasses the analyses that can generalize and transcend existing knowledge. Although not directly analogous, the numerical models/theory contrast brings to mind our dichotomous scheme for categorizing models in Chapter 1, which included "correlative" versus "explanatory" (Section 1.2.4). Correlative models are developed primarily to describe and summarize a set of relationships, without regard for the appropriate representation of processes or mechanisms that operate in the real system. The goal is prediction, not explanation. Explanatory models are developed primarily to represent the internal dynamics of the system-of-interest appropriately. The goal is explanation through representation of the causal mechanisms underlying system behavior. We hastened to add that solely explanatory models and solely correlative models form two ends of a continuum, and classification relative to this dichotomy is based more on the intent of the modeler than on the structure of the model. A model that we view as explanatory at one level of detail we might view as correlative at a finer level of detail.

The numerical models that Ågren and Bosatta (1990) thought were analogous to construction engineering were the huge detailed ecosystem models of the type developed during the 1970s as part of the International Biological Program. The "legacy" models as well as the more recent "integrated" models to which we referred in a previous chapter (Chapter 13, Section 13.2) exemplify these types of models. The science that these authors thought was analogous to art was embodied in the method used in classical (nonquantum) physics. The point of departure for classical physicists is a mental picture of the system-of-interest at a high level of abstraction. Mathematical deduction is the main powerful tool by means of which new and, often, unexpected relationships can be derived between the variables of the theory-model complex. Art, within this context, enters at two points in the process: (a) when choosing the model and (b) when, in the process of deduction, making the "right" approximations.

Thus, Ågren and Bosatta (1990) conclude that the art in science is related intimately to the level of abstraction. We agree, with the caveat that, from a general systems perspective, our perception of a "level of abstraction" is itself abstracted from a continuum of hierarchically nested systems in which each system is a functional component of a larger system and is composed of functional components consisting of smaller systems. All systems are connected directly or indirectly, thus system boundaries are arbitrary and permeable. We recall the "Powers of Ten" film illustrating the hierarchical relatedness of systems that encompass vastly different spatial areas and whose behavior unfolds over vastly different time scales, to which we referred in Chapter 1 (Section 1.4.3). Thus, we would suggest that no matter how complex a final model becomes, regardless of its classification as numerical or theoretical, as correlative or explanatory, it will have had its origins in an artful abstraction of a particular system-of-interest from among all the possible systems in the universe.

Our suggestion that all models begin as an artful abstraction of reality brings to mind our discussion in Chapter 13 of the potential dangers of model integration (Section 13.4). Recall that we defined the process of model integration as the process of integrating (quantitatively coupling) models that originally were developed independently of one another. We argued, based on some of the most basic properties comprising a General Systems Theory of open systems (Von Bertalanffy, 1955; Litterer, 1969, cited in Skyttner, 2005—the book *General Systems Theory*), that it is theoretically impossible to integrate previously built, stand-alone models in a manner such that we can understand them. Our intent was not to suggest that we should not integrate existing models, but to emphasize the importance of always keeping in mind the necessarily arbitrary nature of our model selection process, and that we should judge the appropriateness of each component model to be integrated with reference to the specific objectives of our integrated model. The potential danger, rephrased within the context of the present discussion, is that we risk losing the artfulness involved in the development of the original component models by approaching the model integration process exclusively as a software engineering problem. The preintegration assessment phase of the model integration process (Section 13.3.1) specifically emphasizes the importance of appropriate (artful) integration of the science domains represented in the component models. Thus, ideally, the contextual and semantic problems associated integrating the science domains would be resolved artfully during the preintegration assessment phase. However, to date, the preintegration phase of most integrated modeling projects has not been described well (Belete et al., 2017). We would suggest this most likely is due to the inability to assemble all of the original "artists" to discuss/debate the pertinent scientific issues face-to-face (or at least "Skype-to-Skype").

Pondering further our suggestion that all models begin as an artful abstraction of reality, and considering that Ågren and Bosatta (1990) thought that art was embodied in the elegantly abstract models and powerful mathematical deduction used by classical physicists, somewhat naturally leads us to the idea of "physic envy." Physics envy has haunted not only ecologists, but also economists and sociologists, for decades. Egler (1986) wrote a provocative commentary titled "Physics envy in ecology" in which he argued that the "envy" is borrowed by analogy from the field of psychiatry, and refers to a syndrome in which the methods, techniques, standards, theories, concepts, and paradigms are appropriated and emulated, but have not succeeded in producing the equivalent desired robust results. More recently, Turchin (2001) noted that there remains a widespread opinion among ecologists that ecology lacks general laws and is not a predictive (not a "hard") science. Hence, like many scientists who are not physicists, ecologists have been unable to resist unfavorable comparisons between their science and physics. However, although the behavior of certain physical systems, such as the solar system, can be predicted by laws of classical mechanics much better than ecologists can predict the behavior of any real ecological system, the behavior of most real-world physical systems is much harder to predict than planetary motions, much harder to predict than is portrayed by ecologists suffering from physics-envy. Turchin (2001) further suggests that an important aspect of Newton's genius was in selecting a perfect system, planetary motions, to which to apply his theory. By contrast, ecologists insist on asking really tough and messy questions, such as how many species should comprise an ecological community.

We would suggest that the "envy" in physics envy among ecological modelers, to the extent that it exists, results in large part from the artistry with which physicists choose their systems-of-interest and formulate their conceptual models. However, part of this envy may also come from a misunderstanding of what physicists require of a "law." For example, the requirement of accurate predictions. Respected astrophysicists make theoretical predictions about yet unobserved phenomena, which are impossible to test with manipulative experiments. Physicists have achieved a complete understanding of the laws of fluid dynamics that govern atmospheric movements, yet they cannot predict the weather accurately for more than a few days in advance. It seems to us that ecological modelers often find themselves in situations that are not as dissimilar to those encountered by astrophysicists and experts in fluid dynamics as we might think. Whether or not we exhibit the same degree of artfulness in formulating our conceptual models and defining our requirements for what constitutes a law perhaps merits further reflection.

Recalling our reflections in Chapter 10 related to conceptual validation of models (Section 10.4.1), we note that, like astrophysicists, our inability to test model predictions with manipulative experiments does not in and of itself negate the usefulness of the model. Conceptual validity (following Rykiel, 1996) means that the theories and assumptions underlying the conceptual model are justifiable, and that the model representation of system structure and causal relationships are reasonable for the intended use of the model. Conceptual validity is considered as acceptable justification of the scientific content of the model, but does not guarantee that the model will make accurate predictions. Because the model necessarily is an abstraction of the real system, many components and processes must be left out for the purposes of addressing a particular problem. The amount of knowledge we gain from our model is, indeed, related to model complexity, but the form of the relationship is not unique, nor is it monotonic (see Fig. 11.3). At some point, the benefits to performance of our models that are associated with explicitly representing more system components will be outweighed by the detriments associated with the cumulative effect of errors inherent in estimation of an increasing number of model parameters. An important aspect of the art of modeling is knowing how many and which components and processes to leave out of our models.

Recalling our reflections in Chapter 11 related to what ecological modelers should expect from their models (Section 11.4), we note that, like experts in fluid dynamics, we should not expect to be able to predict real system behavior for more than a few time steps in advance (at best!). In part, this is due, to state the obvious, to the fact that real-world problems are embedded in complex systems. Even the real-world incarnation of the proverbial falling apple is a complex, albeit completely physical, system (assuming a human or some other member of the biota does not interfere with the apple before it hits the ground). Gusts of wind and colliding rain drops are not included in Newton's model (Newton's law), yet virtually no one would claim that the existence of wind and rain negates Newton's law or renders its practical application useless. Thus, regarding the relationship between ecological models and ecological laws, we might pose the simple question, paraphrased after Murray (1992); how exact do ecologists require the correspondence between theory and fact to be before we admit that the theory tells us something useful about an ecological problem?

In this regard, Turchin (2001) has provided a logical argument suggesting that the Malthusian "law" of exponential population growth, which is widely viewed as the first principle of population dynamics (Ginzburg, 1986; Brown, 1997; Berryman, 1999), is directly analogous to the law of inertia (Newton's first law). First, both the law of inertia and the "law" of exponential growth specify the state of the system-of-interest in the absence of any influences acting upon it. The law of inertia states how a body will move in the absence of forces exerted on it. The exponential "law" specifies how a population will grow/decline in the absence of systematic changes in the environmental factors influencing reproduction and mortality. Second, the actions of the law of inertia and the exponential "law" in real life are obscured by the complexities that characterize real-life motions of bodies and population fluctuations, respectively. As a result, neither statement can be subjected to a direct empirical test. Just as we cannot observe a body upon which no forces are acting, we cannot observe a population growing exponentially (at least, not for long) because we cannot indefinitely keep its environment stationary. Thus, Turchin (2001) would argue that ecologists should remove the quotation marks and accept the law of exponential population growth as being on an equal footing with the law of inertia. We agree, and would suggest that perhaps ecological modelers have, in fact, exhibited the same degree of artfulness in formulating some of our conceptual models as have physicists. However, we continue to be blinded to that fact by the seemingly irresistible need to equate the inclusion of complexity explicitly in our models with the inability to identify underlying general laws of system behavior.

Admittedly, there is an additional source of complexity in ecological systems that is not found in purely physical systems. Systems with living components are thermodynamically open, dissipative systems which are maintained far from (thermodynamic) equilibrium by a continual input of energy. Decades ago, in response to an admonition that biologists ought to think more like physicists if they want to develop general theories (Murray, 1992), Quenette and Gerard (1993) suggested that biologists should not think like physicists precisely because we were dealing with thermodynamically open, dissipative systems, far from equilibrium. These are systems whose existence and maintenance require a continuous input of energy, and whose function is to dissipate that energy (lose that energy to their environment). Within this theoretical context, a system that is maintained in a far-from-equilibrium condition by an energy flow can display new self-organized structures which appear beyond a threshold of instability. For example, if any perturbation from the environment leads a nonlinear dynamic system to an instability point (a point of bifurcation), it can evolve new attractors corresponding to new stable functioning regimes. (This is a fancy way of saying that there are limits to system resilience, a concept that often is depicted in basic ecology texts by a simple physical example, as we did in Chapter 1, Section 1.4.4, Fig. 1.5.) Quenette and Gerard (1993), among others, foresaw that the (then) newly developing theory of dynamic systems would give rise to a new way of thinking about the dynamics of biological phenomena observed at different levels of integration, including molecular (Goldbeter, 1991), organismal (Waddington, 1975), and population (May and Oster, 1976).

So, does present day systems theory, in addition to providing the framework for thinking about the dynamics of biological phenomena, also provide insight into the art of modeling and the definition of ecological laws? Our answer to this question, not surprisingly, is yes. And, in a way, the rationale for our affirmative answer brings us full circle to Chapter 1 of this book, in which we presented the basic characteristics of systems (Section 1.4) and of systems models (Section 1.4.6). A system is more than the sum of its parts, system structure is the source system behavior, and systems are hierarchical, resilient (within limits), and self-organizing. All systems possess these five characteristics. They are laws. We noted that not all of our models of systems, not all of the systems-of-interest that we define to address our questions, necessarily will exhibit each of these five system characteristics. By definition, all will exhibit the first two. Deciding which, if any, of the remaining system characteristics to include explicitly in any given model is part of the art of modeling. The usefulness of our models depends directly on the artfulness of our necessarily arbitrary decisions with regard to bounding the system-of-interest in space and time, and with regard to the level of detail with which we choose to represent system components.

We want to close this chapter and this book with brief summaries of three, perhaps seemingly disparate and randomly chosen, articles. The first is titled "The art and science of an analytic potential" (Brenner, 2000). The second is titled "The edges of understanding" (Lander, 2010). The third is a commentary/book review titled "Superforecasting" (Brown, 2015). Our intent is to leave you, the readers, with some food for thought, something to ponder, to reflect upon, with regard to the "art" in the art and science of modeling complex systems, ecological or otherwise.

Brenner (2000), in the introduction to his article, explains that analytic potential energy functions, which sometimes are referred to as empirical or classical potentials, are simplified mathematical expressions that attempt to model interatomic forces arising from the quantum mechanical interactions of electrons and nuclei. Their use usually is necessitated either by the desire to model systems with sizes and/or timescales that exceed the available computing resources required for quantum calculations, or to gain qualitative insight into things like bonding preferences that may not be immediately obvious from the results of numerical calculations. To be effective, an analytic potential energy function should be (1) flexible enough to accommodate a wide a range of fitting data, (2) able to reproduce accurately an appropriate fitting data base, (3) transferable enough to describe at least qualitatively, if not with quantitative accuracy, structures not included in a fitting data base, and (4) computationally efficient. Often, criteria 1 and 2 are emphasized in the development of an analytic potential, with the assumption that these will lead to transferability. Unfortunately, especially with ad hoc functional forms, as more arbitrary fitting parameters are added, functions often lose transferability. Analytic potential functions with the highest degree of transferability are those based on sound quantum-mechanical bonding principles, and not necessarily those with the most parameters. Thus, the trick to producing an effective analytic potential energy function is balancing sound functional forms with any necessary empirical parameter fitting. This process often requires some level of intuitive chemical insight, considerable trial-and-error, and significant tenacity. Because there is no rigorous methodology in this process with which all systems can be fit, it is often referred to as an "art."

Lander (2010) opens his article with the provocative suggestion that the hairball is replacing the double helix as the dominant icon of molecular systems biology. Although not as aesthetically pleasing as the double helix of 20th century molecular biology, the hairball is iconic because it captures succinctly the distinctive flavor of the present, postmodern era of systems biology. The "hairball" icon represents a network diagram in which the vertices or "nodes" represent genes, proteins, or metabolites, and the connectors or "edges" represent interactions between pairs of these components (see Fig. 14.3). Both molecular and systems biologists construct their view of biology based on knowledge of biological components (nodes) and knowledge of the relationships among those components (edges). However, molecular biologists focus primarily on discovering nodes and linking them through edges that stand for qualitative causal relationships ("gene a turns on gene b"), whereas systems biologists focus primarily on investigation of the networks themselves, the behaviors of which tend to be dominated by the quantitative details of their edges more than by the physical nature of their nodes.

Lander (2010) continues his hairball analogy to make the point that systems biology is more than the exploitation of high-throughput methods to gather vast amounts of data about genomes, epigenomes, transcriptomes, proteomes, metabolomes, phenomes, and the like. That is, systems biology is more than simply discovering nodes and edges. In driving home this point, Lander assigns different meanings to the nodes and the edges of the hairball. Now, the nodes represent knowledge in the form of individual sets of data about the biological world, including facts, observations, structures, behaviors, and so on. The edges now represent relationships, or connections, between bodies of knowledge. For example, what we know about the cell cycle and what we know about circadian behaviors such as sleep-wake cycles may be connected by virtue of the fact that both phenomena are built upon autonomous oscillators with external inputs. When we appreciate that two or more seemingly very different phenomena can be treated as similar in some way, we feel that we have accomplished something. We call this feeling "understanding." There is an important distinction between knowledge and understanding. Factual discovery, whether in the biological sciences or any other enterprise, does not in and of itself constitute understanding.

So how can we create understanding out of many validated bits of knowledge? Lander (2010) notes that there are many phrases that describe the action of replacing the messy with the simple to promote understanding: "creating an abstraction," "generalizing," and "distilling a concept" come to mind, but the phrase he finds most evocative is, "building a model." When we understand the cell cycle as a regulated oscillator, or metabolic networks as systems for optimizing growth, we are building simple models of complex processes. Models do not arise by logical inference from data, they are acts of human creation. Any large set of data can be modeled in a large (approaching infinite) number of ways. Our reasons for choosing one way over another are not to be found in the data themselves, but rather in our ideas about how a model will help us connect the data to other knowledge to increase our understanding of the real-world processes that generated the patterns that we see in those data.

Brown's (2015) "Superforecasting" is a commentary/book review in which he reflects on portions of the book titled "Superforecasting: The art and science of prediction" (Tetlock and Gardner, 2016). Among other things, the book describes an experiment, a prediction tournament, that pitted amateur forecasters recruited by Tetlock and his graduate students against four other academic teams. The other teams used methods such as prediction trading markets and machine learning algorithms, as well as consultations with government analysts (who, among other advantages, had access to classified information). The questions were ones of interest to national policy makers, such as: "Who will be in control of Yemen at the end of 2015?" with forecasters asked to assign probabilities to answers such as: "President Hadi will be restored to power," or "The civil war will be continuing." Tetlock's teams did much better than the other teams, and the scale of victory was impressive, 30% better, as measured by the Brier score. In a binary question, that translates to assigning 20% less probability, on average, to the wrong answer. Based on the results of this experiment, together with some insights into the behavioral tendencies of the victorious amateurs, Tetlock and Gardner (2016) listed "ten commandments for aspiring superforecasters:" (1) Triage. (2) Break seemingly intractable problems into tractable subproblems. (3) Strike the right balance between inside and outside views. (4) Strike the right balance between under- and over-reacting to evidence. (5) Look for the clashing causal factors at work in each problem. (6) Strive to distinguish as many degrees of doubt as the problem permits, but no more. (7) Strike the right balance between under- and overconfidence, between prudence and decisiveness. (8) Look for the errors behind your mistakes but beware of rearview-mirror hindsight biases. (9) Bring out the best in others and let others bring out the best in you. (10) Master the error-balancing bicycle. Brown (2015) offered some thoughts regarding each of these 10 commandments, which merit reflection, but which we will not summarize here. Admittedly, some of the commandments are rather cryptic without further explanation. But, all the better to ponder.

Thus, we leave you with thoughts of modeling interatomic forces arising from the quantum mechanical interactions of electrons and nuclei, the hairball replacing the double helix as the dominant icon of molecular systems biology, and the 10 commandments for aspiring superforecasters. We hope that reflecting upon and attempting to make connections among these seemingly disparate topics from a systems perspective will increase your appreciation for the "art" in the art and science of modeling ecological systems.

References

Ågren, G.I., Bosatta, E., 1990. Theory and model or art and technology in ecology. Ecol. Model. 50 (1–3), 213–220.

Ammann, P., Offutt, J., 2008. Introduction to Software Testing. Cambridge University Press.

Argent, R.M., 2004. An overview of model integration for environmental applications—components, frameworks and semantics. Environ. Model Softw. 19 (3), 219–234.

Ascough II, J., Maier, H., et al., 2008. Future research challenges for incorporation of uncertainty in environmental and ecological decision-making. Ecol. Model. 219 (3–4), 383–399.

Augusiak, J., Van den Brink, P.J., et al., 2014. Merging validation and evaluation of ecological models to 'evaludation': a review of terminology and a practical approach. Ecol. Model. 280, 117–128.

Baker, W.L., 1989. A review of models of landscape change. Landsc. Ecol. 2 (2), 111–133.

Barlas, Y., 1996. Formal aspects of model validity and validation in system dynamics. Syst. Dyn. Rev. 12 (3), 183–210.

Barlas, Y., Carpenter, S., 1990. Philosophical roots of model validation: two paradigms. Syst. Dyn. Rev. 6 (2), 148–166.

Bart, J., 1995. Acceptance criteria for using individual-based models to make management decisions. Ecol. Appl. 5 (2), 411–420.

Bas van Fraassen, C., 1980. The Scientific Image. Oxford University Press, New York, NY.

Beck, M., Ravetz, J., et al., 1997. On the problem of model validation for predictive exposure assessments. Stoch. Hydrol. Hydraul. 11 (3), 229–254.

Belete, G.F., Voinov, A., 2016. Exploring temporal and functional synchronization in integrating models: a sensitivity analysis. Comput. Geosci. 90, 162–171.

Belete, G.F., Voinov, A., et al., 2017. An overview of the model integration process: from pre-integration assessment to testing. Environ. Model Softw. 87, 49–63.

Berner, S., Weber, R., et al., 2005. Observations and lessons learned from automated testing. In: Proceedings of the 27th International Conference on Software Engineering. ACM.

Berryman, A.A., 1999. Principles of Population Dynamics and their Application. Stanley Thornes Publishers Ltd, Cheltenham, UK.

Bertolino, A., 2007. Software testing research: achievements, challenges, dreams. In: 2007 Future of Software Engineering. IEEE Computer Society.

Booth, N.L., Everman, E.J., et al., 2011. A web-based decision support system for assessing regional water-quality conditions and management actions 1. J. Am. Water Resour. Assoc. 47 (5), 1136–1150.

Brenner, D.W., 2000. The art and science of an analytic potential. Phys. Status Solidi B 217 (1), 23–40.

Brooking, C., Hunter, J., 2013. Providing online access to hydrological model simulations through interactive geospatial animations. Environ. Model Softw. 43, 163–168.

Brown, James H., 1997. An Ecological Perspective on the Challenge of Complexity. EcoEssay Series Number 1. National Center for Ecological Analysis and Synthesis, Santa Barbara, CA.

Brown, A., 2015. Superforecasting. Wilmott 2015 (78), 12–15.

Bruggeman, J., Bolding, K., 2014. A general framework for aquatic biogeochemical models. Environ. Model Softw. 61, 249–265.

Butterfield, M.L., Pearlman, J.S., et al., 2008. A system-of-systems engineering GEOSS: architectural approach. IEEE Syst. J. 2 (3), 321–332.

Cartwright, S.J., Bowgen, K.M., et al., 2016. Communicating complex ecological models to non-scientist end users. Ecol. Model. 338, 51–59.

Caswell, H., 1976. The validation problem. In: Patten, B. (Ed.), Systems Analysis and Simulation in Ecology. vol. 4. Academic Press, New York, NY, pp. 313–325.

Caswell, H., 1988. Theory and models in ecology: a different perspective. Ecol. Model. 43 (1–2), 33–44.

Caswell, H., 2006. Matrix Population Models: Construction, Analysis, and Interpretation. Sinauer Associates Inc, Sunderland, MA.

Cess, R.D., Potter, G., et al., 1990. Intercomparison and interpretation of climate feedback processes in 19 atmospheric general circulation models. J. Geophys. Res. Atmos. 95 (D10), 16601–16615.

Chang, H., Lee, K., 2004. Applying web services and design patterns to modeling and simulating real-world systems. In: International Conference on AI, Simulation, and Planning in High Autonomy Systems. Springer.

Chesson, J., 1978. Measuring preference in selective predation. Ecology 59 (2), 211–215.

Clark, J.S., 2005. Why environmental scientists are becoming Bayesians. Ecol. Lett. 8 (1), 2–14.

Cochran, W., Cox, G., 1957. Experimental Designs. John Wiley & Sons, New York, NY.

Curry, G.L., Feldman, R.M., et al., 1989. Discrete Simulation: Fundamentals and Microcomputer Support. Holden-Day, Inc, San Francisco, CA.

Dubois, G., Schulz, M., et al., 2013. eHabitat, a multi-purpose Web Processing Service for ecological modeling. Environ. Model Softw. 41, 123–133.

Egler, F.E., 1986. "Physics envy" in ecology. Bull. Ecol. Soc. Am. 67 (3), 233–235.

Erl, T., 2008. SOA Principles of Service Design. Prentice Hall, Upper Saddle River.

Fagerström, T., 1987. On theory, data and mathematics in ecology. Oikos 50 (2), 258–261.

Fath, B.D., Patten, B.C., 1999. Review of the foundations of network environ analysis. Ecosystems 2 (2), 167–179.

Fath, B.D., Scharler, U.M., et al., 2007. Ecological network analysis: network construction. Ecol. Model. 208 (1), 49–55.

Fishman, G.S., Kiviat, P.J., 1968. The statistics of discrete-event simulation. Simulation 10 (4), 185–195.

Fleming, R.A., Shoemaker, C.A., 1992. Evaluating models for spruce budworm-forest management: comparing output with regional field data. Ecol. Appl. 2 (4), 460–477.

Ford, F.A., 1999. Modeling the Environment: An Introduction to System Dynamics Models of Environmental Systems. Island Press, Washington, DC.

Forrester, J.W., 1961. Industrial Dynamics. Pegasus Communications, Waltham, MA.

Frakes, W.B., Kang, K., 2005. Software reuse research: status and future. IEEE Trans. Softw. Eng. 31 (7), 529–536.

French, N., Sauer, R.H., 1974. Phenological studies and modeling in grasslands. In: Lieth, H. (Ed.), Phenology and Seasonality Modeling. Springer-Verlag, Berlin, Heidelberg, pp. 227–236.

French, N., Grant, W., et al., 1976. Small mammal energetics in grassland ecosystems. Ecol. Monogr. 46 (2), 201–220.

Gass, S.I., 1983. Decision-aiding models: validation, assessment, and related issues for policy analysis. Oper. Res. 31 (4), 603–631.

Geller, G.N., Turner, W., 2007. The model web: a concept for ecological forecasting. In: IEEE International Geoscience and Remote Sensing Symposium, 2007. IGARSS 2007. IEEE.

Gilman, M., Hails, R., 1997. An Introduction to Ecological Modelling: Putting Practice Into Theory. Wiley-Blackwell, Hoboken, NJ.

Ginzburg, L.R., 1986. The theory of population dynamics: I. Back to first principles. J. Theor. Biol. 122 (4), 385–399.

Goldbeter, A., 1991. A minimal cascade model for the mitotic oscillator involving cyclin and cdc2 kinase. Proc. Natl. Acad. Sci. 88 (20), 9107–9111.

Goodall, J.L., Robinson, B.F., et al., 2011. Modeling water resource systems using a service-oriented computing paradigm. Environ. Model Softw. 26 (5), 573–582.

Grant, W.E., 1986. Systems Analysis and Simulation in Wildlife and Fisheries Sciencies. John Wiley, New York, NY.

Grant, W.E., 1998. Ecology and natural resource management: Reflections from a systems perspective. Ecol. Model. 108 (1–3), 67–76.

Grant, W.E., Swannack, T.M., 2008. Ecological Modeling: A Common-Sense Approach to Theory and Practice. Blackwell Publishing, Hoboken, NJ.

Grant, W.E., Pedersen, E.K., et al., 1997. Ecology and Natural Resource Management: Systems Analysis and Simulation. John Wiley & Sons, New York, NY.

Griffies, S.M., 2012. Elements of the modular ocean model (MOM). GFDL Ocean Group Tech. Rep. 7, p. 620.

Grimm, V., Railsback, S.F., 2005. Individual-Based Modeling and Ecology. Princeton University Press, Princeton, NJ.

Grimm, V., Railsback, S.F., 2012. Pattern-oriented modelling: a 'multi-scope' for predictive systems ecology. Philos. Trans. Royal Soc. B Biol. Sci. 367 (1586), 298–310.

Grimm, V., Berger, U., et al., 2006. A standard protocol for describing individual-based and agent-based models. Ecol. Model. 198 (1–2), 115–126.

Hastings, N., Peacock, J., 1975. Statistical Distributions: A Handbook for Students and Practitioners. Wiley.

Holling, C.S., 1978. Adaptive Environmental Assessment and Management. John Wiley & Sons, Chichester, UK.

Holzworth, D.P., Snow, V., et al., 2015. Agricultural production systems modelling and software: current status and future prospects. Environ. Model Softw. 72, 276–286.

Hoover, S.V., Perry, R.F., 1989. Simulation: A Problem-Solving Approach. Addison-Wesley, Reading, MA.

Hughes, J.D., Langevin, C.D., Banta, E.R., 2017. Documentation for the MODFLOW 6 Framework (No. 6-A57). US Geological Survey.

Hunt, H.W., 1977. A simulation model for decomposition in grasslands. Ecology 58 (3), 469–484.

Innis, G., 1979. A spiral approach to ecosystem simulation. In: Innis, G.S., O'Neill, R.V. (Eds.), Systems Analysis of Ecosystem. International Cooperative Publishing House, Burtonsville, MD.

ISO/IEC, 2011. IEC/IEEE Systems and Software Engineering: Architecture Description. ISO/IEC/IEEE 42010: 2011 (E) (Revision of ISO/IEC 42010: 2007 and IEEE Std 1471-2000).

Jakeman, A.J., Letcher, R.A., et al., 2006. Ten iterative steps in development and evaluation of environmental models. Environ. Model Softw. 21 (5), 602–614.

Johnson, N.L., Kotz, S., 1969. Distributions in Statistics. Wiley, New York, NY.

Jørgensen, S.E., Bendoricchio, G., 2001. Fundamentals of Ecological Modelling. Elsevier, Amsterdam, The Netherlands.

Kitching, G.N., 1983. Rethinking Socialism: A Theory for a Better Practice. Methuen, London, UK.

Kleiber, M., 1947. Body size and metabolic rate. Physiol. Rev. 27 (4), 511–541.

Kuhn, T., 1962. The Structure of Scientific Revolutions. Chicago University Press, Chicago, IL.

Lander, A.D., 2010. The edges of understanding. BMC Biol. 8 (1), 40.

Langevin, C.D., Hughes, J.D., Banta, E.R., Niswonger, R.G., Panday, S., Provost, A.M., 2017. Documentation for the MODFLOW 6 Groundwater Flow Model (No. 6-A55). US Geological Survey.

Laniak, G.F., Olchin, G., et al., 2013. Integrated environmental modeling: a vision and roadmap for the future. Environ. Model Softw. 39, 3–23.

Lefkovitch, L.P., 1965. The study of population growth in organisms grouped by stages. Biometrics 21, 1–18.

Leslie, P.H., 1945. On the use of matrices in certain population mathematics. Biometrika 33 (3), 183–212.

Lewis, R.O., 1992. Independent Verification and Validation: A Life Cycle Engineering Process for Quality Software. John Wiley & Sons, New York, NY.

Litterer, J.A., 1969. Organizations: Systems, Control and Adaptations. Wiley, Hoboken, NJ.

Loehle, C., 1987. Hypothesis testing in ecology: psychological aspects and the importance of theory maturation. Quart. Rev. Biol. 62 (4), 397–409.

Luo, L., 2001. Software Testing Techniques. Institute for Software Research International, Carnegie Mellon University, Pittsburgh, PA, p. 19.

Madni, A.M., Sievers, M., 2014. Systems integration: key perspectives, experiences, and challenges. Syst. Eng. 17 (1), 37–51.

Maier, H., Ascough II, J., et al., 2008. Chapter 5: Uncertainty in environmental decision making: issues, challenges and future directions. In: Jakeman, A.J., Voinov, A.A., Rizzoli, A.E., Chen, S.H. (Eds.), Developments in Integrated Environmental Assessment. vol. 3. Elsevier, pp. 69–85.

Mankin, J., O'Neill, R., et al., 1977. The importance of validation in ecosystem analysis. In: Innis, G.S. (Ed.), New Directions in the Analysis of Ecological Systems. vol. 1. The Society for Computer Simulation, La Jolla, CA, pp. 63–71.

May, R.M., Oster, G.F., 1976. Bifurcations and dynamic complexity in simple ecological models. Am. Nat. 110 (974), 573–599.

Mayer, D., Butler, D., 1993. Statistical validation. Ecol. Model. 68 (1–2), 21–32.

McDonald, M.G., Harbaugh, A.W., 2003. The history of MODFLOW. Ground Water 41 (2), 280–283.

Meadows, D.H., 2008. Thinking in Systems: A Primer. Chelsea Green Publishing, White River Junction, VT.

Mentis, M., 1988. Hypothetico-deductive and inductive approaches in ecology. Funct. Ecol. 2 (1), 5–14.

Moore, R.V., Tindall, C.I., 2005. An overview of the open modelling interface and environment (the OpenMI). Environ. Sci. Policy 8 (3), 279–286.

Murray, B.G., 1992. Research methods in physics and biology. Oikos 64 (3), 594–596.

Myers, G.J., Sandler, C., et al., 2011. The Art of Software Testing. John Wiley & Sons.

Nativi, S., Mazzetti, P., et al., 2013. Environmental model access and interoperability: the GEO Model Web initiative. Environ. Model Softw. 39, 214–228.

Oreskes, N., Belitz, K., 2001. Philosophical issues in model assessment. In: Anderson, M.G., Bates, P.D. (Eds.), Model Validation: Perspectives in Hydrological Science. Wiley, pp. 23–41.

Oreskes, N., Shrader-Frechette, K., et al., 1994. Verification, validation, and confirmation of numerical models in the earth sciences. Science 263 (5147), 641–646.

Ott, R.L., 1984. An Introduction to Statistical Methods and Data Analysis. Duxbury Press, Boston, MA.

Ott, R.L., Longnecker, M., 2001. An Introduction to Statistical Methods and Data Analysis. Duxbury Press, Belmont, CA.

Patten, B.C., 1985. Energy cycling in the ecosystem. Ecol. Model. 28 (1–2), 1–71.

Peckham, S.D., Hutton, E.W., et al., 2013. A component-based approach to integrated modeling in the geosciences: the design of CSDMS. Comput. Geosci. 53, 3–12.

Peterson, M.N., Peterson, M.J., et al., 2005. Conservation and the myth of consensus. Conserv. Biol. 19 (3), 762–767.

Popper, K.R., 1959. The Logic of Scientific Discovery. Routledge, Abingdon, UK.

Power, M., 1993. The predictive validation of ecological and environmental models. Ecol. Model. 68 (1–2), 33–50.

Proulx, S.R., Promislow, D.E., et al., 2005. Network thinking in ecology and evolution. Trends Ecol. Evol. 20 (6), 345–353.

Quenette, P.Y., Gerard, J.F., 1993. Why biologists do not think like Newtonian physicists. Oikos 68 (2), 361–363.

Ramamoorthy, C., Chandra, C., et al., 1992. Proceedings of the Second International Conference on Systems integration: Problems and Approaches (ICSI'92). IEEE.

Ramamurthy, K., King, W.R., Premkumar, G., 1992. User characteristics – DSS effectiveness linkage: an empirical assessment. Int. J. Man-Mach. Stud. 36 (3), 469–505.

Refsgaard, J.C., Henriksen, H.J., 2004. Modelling guidelines—terminology and guiding principles. Adv. Water Resour. 27 (1), 71–82.

Refsgaard, J.C., Henriksen, H.J., et al., 2005. Quality assurance in model based water management – review of existing practice and outline of new approaches. Environ. Model Softw. 20 (10), 1201–1215.

Regan, H.M., Colyvan, M., et al., 2002. A taxonomy and treatment of uncertainty for ecology and conservation biology. Ecol. Appl. 12 (2), 618–628.

Renner, S., 2001. A community of interest approach to data interoperability. In: Federal Database Colloquium. CiteSeerX, San Diego, CA.

Rykiel Jr., E.J., 1984. Modeling agroecosystems: lessons from ecology. In: Lowrance, R., Stinner, B., House, G. (Eds.), Agricultural Ecosystems: Unifying Concepts. Wiley, New York, NY, pp. 157–178.

Rykiel Jr., E.J., 1996. Testing ecological models: the meaning of validation. Ecol. Model. 90 (3), 229–244.

Sargent, R.G., 1984. A tutorial on verification and validation of simulation models. In: Proceedings of the 16th Conference on Winter Simulation. IEEE Press.

Sargent, R.G., 2010. Verification and validation of simulation models. In: Proceedings of the 2010 Winter Simulation Conference. IEEE.

Sargent, R.G., 2013. Verification and validation of simulation models. J. Simul. 7 (1), 12–24.

Schlesinger, S., 1979. Terminology for model credibility. Simulation 32 (3), 103–104.

Skyttner, L., 2005. General Systems Theory: Problems, Perspectives, Practice. World Scientific.

Smith, F.E., 1973. Analysis of ecosystems. In: Reichle, D.E. (Ed.), Analysis of Temperate Forest Ecosystems. Springer-Verlag, New York, NY, pp. 1–46.

Snedecor, G.W., Cochran, W.G., 1967. Statistical Methods. Iowa State University Press, Ames, IA.

Sokal, R., Rohlf, F., 1969. Biometry: The Principles and Practice of Statistics in Biological Research. W. H. Freeman and Company, San Francisco, CA.

Spain, J.D., 1982. BASIC Microcomputer Models in Biology. Addison-Wesley, London, UK.

Starfield, A.M., Bleloch, A.L., 1986. Building models for conservation and wildlife management. Macmillan.

Stein, A., Draxler, R.R., et al., 2015. NOAA's HYSPLIT atmospheric transport and dispersion modeling system. Bull. Am. Meteorol. Soc. 96 (12), 2059–2077.

Steinhorst, R., 1979. Parameter identifiability, validation, and sensitivity analysis of large system models. In: Innis, G.S., O'Neill, R.V. (Eds.), Systems Analysis of Ecosystems. International Cooperative Publishing House, Burtonsville, MD, pp. 33–58.

Suppe, F., 1989. The Semantic Conception of Theories and Scientific Realism. University of Illinois Press, Urbana, IL.

Tetlock, P.E., Gardner, D., 2016. Superforecasting: The Art and Science of Prediction. Random House, New York, NY.

Thagard, P., 1988. Computational Philosophy of Science. MIT Press, Cambridge, MA.

Topping, C.J., Dalkvist, T., et al., 2012. Post-hoc pattern-oriented testing and tuning of an existing large model: lessons from the field vole. PLoS One 7 (9), e45872.

Trusted, J., 1979. The Logic of Scientific Inference: An Introduction. Macmillan International Higher Education, New York, NY.

Turchin, P., 2001. Does population ecology have general laws? Oikos 94 (1), 17–26.

Van den Belt, M., 2004. Mediated Modeling: A System Dynamics Approach to Environmental Consensus Building. Island Press, Washington, DC.

van der Sluijs, J.P., 1997. Anchoring Amid Uncertainty: On the Management of Uncertainties in Risk Assessment of Antropogenic Climate Change. Universiteit Utrecht.

van der Sluijs, J., van Eijndhoven, J., et al., 1998. Anchoring devices in science for policy: the case of consensus around climate sensitivity. Soc. Stud. Sci. 28 (2), 291–323.

Van Dyne, G.M., 1969. Grasslands Management, Research, and Training Viewed in a Systems Context. Colorado State University, Fort Collins, CO.

Van Ittersum, M.K., Ewert, F., et al., 2008. Integrated assessment of agricultural systems – a component-based framework for the European Union (SEAMLESS). Agric. Syst. 96 (1–3), 150–165.

Voinov, A., Cerco, C., 2010. Model integration and the role of data. Environ. Model Softw. 25 (8), 965–969.

Voinov, A., Shugart, H.H., 2013. 'Integronsters', integral and integrated modeling. Environ. Model Softw. 39, 149–158.

Von Bertalanffy, L., 1955. An essay on the relativity of categories. Philos. Sci. 22 (4), 243–263.

Waddington, C., 1975. The Evolution of an Evolutionist. Cornell University Press, Ithaca, NY.

Walker, W.E., Harremoës, P., et al., 2003. Defining uncertainty: a conceptual basis for uncertainty management in model-based decision support. Integr. Assess. 4 (1), 5–17.

Walters, C.J., 1986. Adaptive Management of Renewable Resources. Macmillan Publishers Ltd, New York, NY.

Wang, M., Luttik, R., 2012. Population level risk assessment: practical considerations for evaluation of population models from a risk assessor's perspective. Environ. Sci. Eur. 24 (1), 3.

Weinberg, G.M., 1975. An introduction to general systems thinking. Vol. 304. New York, Wiley.

Zurell, D., Grimm, V., et al., 2012. Uncertainty in predictions of range dynamics: black grouse climbing the Swiss Alps. Ecography 35 (7), 590–603.

Index

Note: Page numbers followed by *f* indicate figures, and *t* indicate tables.